I0053255

Shape Memory Alloys
SMA 2018

Selected, peer reviewed papers from the
International Conference
"Shape memory alloys", August 16-20 2018,
Chelyabinsk, Russia

Edited by
Vasiliy Buchelnikov, Vladimir Sokolovskiy,
Mikhail Zagrebin, Olga Miroshkina

Peer review statement

All papers published in this volume of "Materials Research Proceedings" have been peer
reviewed. The process of peer review was initiated and overseen by the above
proceedings editors. All reviews were conducted by expert referees in accordance to
Materials Research Forum LLC high standards.

Copyright © 2018 by authors

[(cc) BY] Content from this work may be used under the terms of the Creative Commons Attribution 3.0 license. Any further distribution of this work must maintain attribution to the author(s) and the title of the work, journal citation and DOI.

Published under License by **Materials Research Forum LLC**
Millersville, PA 17551, USA

Published as part of the proceedings series
Materials Research Proceedings
Volume 9 (2018)

ISSN 2474-3941 (Print)
ISSN 2474-395X (Online)

ISBN 978-1-64490-000-0 (Print)
ISBN 978-1-64490-001-7 (eBook)

This book contains information obtained from authentic and highly regarded sources. Reasonable efforts have been made to publish reliable data and information, but the author and publisher cannot assume responsibility for the validity of all materials or the consequences of their use. The authors and publishers have attempted to trace the copyright holders of all material reproduced in this publication and apologize to copyright holders if permission to publish in this form has not been obtained. If any copyright material has not been acknowledged please write and let us know so we may rectify in any future reprint.

Distributed worldwide by

Materials Research Forum LLC
105 Springdale Lane
Millersville, PA 17551
USA
http://www.mrforum.com

Manufactured in the United State of America
10 9 8 7 6 5 4 3 2 1

Table of Contents

Structure, Martensitic Transformations and Shape Memory Effects in Alloys

The Theory of Martensitic Transformations and Shape Memory Effect: Modeling and Calculations

Novel Materials: Design, Synthesis, Functional Properties

Preface

The present volume contains selected papers from the 3rd International Conference "Shape memory alloys" (SMA 2018). The conference is organized by Chelyabinsk State University (Chelyabinsk, Russia), Saint Petersburg State University (Saint Petersburg, Russia) and National University of Science and Technology "MISIS" (Moscow, Russia).

The conference continues the tradition of regular seminars and conferences devoted to shape memory alloys which have been held in different cities of the Soviet Union: Kiev (1980, 1991), Voronezh (1982), Tomsk (1985), Novgorod (1989), Kosov (1991), St. Petersburg (1995). The first conference "Shape memory alloys" was previously held in 2014 (Vitebsk, Belarus) and 2016 (Saint Petersburg, Russia). The aim of the conference is to review modern research and development in the field of shape memory alloys and related phenomena: from studying their structure, physical, mechanical, and functional properties to mathematical modeling of the shape memory materials behavior and their application. Also, the satellite seminar "New materials: design, synthesis, functional properties" accompanied the conference. The topics of the conference were as follows:

- Structure, martensitic transformations and shape memory effects in alloys.

- The theory of martensitic transformations and shape memory effect: modeling and calculations.

- Novel materials. The manufacturing technology and application of shape memory alloys.

- Materials design, modeling and calculations of functional properties.

The present volume includes 33 articles, which were presented at the conference. All of the articles were subjected to peer reviewing by two expert referees. The papers selected for this volume depended on their quality and their relevancy to the conference. The volume tends to present to the reader recent advances in the fields of martensitic phase transformations, thermally and induced shape memory effects as well as physic properties accompanying the shape memory effect.

The Organizing Committee is grateful to all of the contributors who made this volume possible. As the guest editors of the volume, we wish to acknowledge all of those who have updated and reviewed the papers submitted to the conference.

Guest editors of SMA 2018 Proceedings
Prof. Vasiliy Buchelnikov, buche@csu.ru
Dr. Vladimir Sokolovskiy, vsokolovsky84@mail.ru
Dr. Mikhail Zagrebin, miczag@mail.ru
Mrs. Olga Miroshkina, miroshkina.on@yandex.ru
Chelyabinsk State University, Chelyabinsk, Russia

Committees

Conference organizers

Chelyabinsk State University (Chelyabinsk, Russia)
Saint Petersburg State University (Saint Petersburg, Russia)
National University of Science and Technology "MISIS" (Moscow, Russia)

Conference Chairs

V.D. Buchelnikov, Chelyabinsk, Russia
S.D. Prokoshkin, Moscow, Russia
N.N. Resnina, Saint Petersburg, Russia

International Advisory Committee

S.P. Belyaev, Saint Petersburg, Russia
V.I. Betekhtin, Saint Petersburg, Russia
V. Brailovski, Montreal, Canada
R.Z. Valiev, Saint Petersburg, Russia
A.E. Volkov, Saint Petersburg, Russia
A.M. Glezer, Moscow, Russia
V.E. Gunther, Tomsk, Russia
M.P. Kaschenko, Ekaterinburg, Russia
Yu.N. Koval, Kiev, Ukraine
S.B. Kustov, Palma, Spain
A.I. Lotkov, Tomsk, Russia
A.A. Movchan, Moscow, Russia
R.R. Mulyukov, Ufa, Russia
V.V. Rubanik Jr., Vitebsk, Belarus
V.A. Plotnikov, Barnaul, Russia
G.S. Firstov, Kiev, Ukraine
V.V. Khovailo, Moscow, Russia
Yu.I. Chumlyakov, Tomsk, Russia

Conference sponsors

The Russian Foundation for Basic Research
Leninsky prospect, 32A, 119991, B-334, GSP-1, Moscow, Russia
Tel: +7495-952-58-47, Fax: +7495-938-19-31
www.rfbr.ru
(Grant # 18-02-20089 G)

Chelyabinsk State University
Bratiev Kashirinykh Str. 129, 454001, Chelyabinsk, Russia
Tel: +7351-799-72-16, Fax: +7351-799-71-25
www.csu.ru
Grant of Scientific Foundation of the Faculty of Physics

Shape Memory Alloys – SMA 2018
Materials Research Proceedings **9** (2018)

Materials Research Forum LLC
doi: http://dx.doi.org/10.21741/9781644900017

Structure, Martensitic Transformations and Shape Memory Effects in Alloys

Shape Memory Alloys – SMA 2018   Materials Research Forum LLC
Materials Research Proceedings **9** (2018) 3-8   doi: http://dx.doi.org/10.21741/9781644900017-1

Technological Features of Wire with a Diameter of 0.5–2.5 mm Production from Ni –Ti-based Shape Memory Alloys

Vladimir A. Andreev[1,2,a*], Vladimir S. Yusupov[1,b], Mikhail M. Perkas[1,c],
Sofya A. Bondareva[3,d], Roman D. Karelin[1,3,e]

[1]A.A. Baikov Institute of Metallurgy and Materials Science, RAS, 49 Leninskiy Prospect, Moscow, 119991, Russia

[2]MATEK-SMA Ltd., 2a-137 Karier Str., Moscow 117449, Russia

[3]National University of Science and Technology "MISiS", 4 Leninskiy Prospect, Moscow, 119049, Russia

[a]andreev.icmateks@gmail.com, [b]yusupov@aport2000.ru, [c]perkas03@yandex.ru, [d]sonya60@rambler.ru, [e]rdkarelin@gmail.com

*corresponding author

Keywords: Shape Memory Alloys, Drawing, Nitinol, Thermomechanical Treatment, Functional Properties, Wire

Abstract. In present work technological process of wire production with a diameter of 0.5–2.5 mm from Ni–Ti-based shape memory alloys by hot and warm drawing, which was developed and applied in the industrial center MATEK-SMA Ltd, was described. Chemical composition of the wire, technical specification of incorporated equipment and temperature-deformational regimes of drawing, that allow obtaining wire, complied with technological conditions No 18.4270–001–16980791–2013, was distinguished. Developed technology allows enhancing production capacity and obtaining wire with improved mechanical and functional properties in comparison with the previously used one.

Introduction

Wire with a diameter of 0.5–2.5 mm from Ni–Ti-based shape memory alloys is widely used as a blank for manufacturing of various devices and functional elements, especially for medical application [1,2]. Chemical composition of the wire depends on the required operational characteristics and vary from 54.6 to 57.0 wt.% Ni. It is well-known that alloy composition in this range of Ni content has a huge influence on the mechanical and functional properties of the material and may decrease the technological deformability [3-5].

Technology of production semi-finished products from Ni–Ti-based shape memory alloys was firstly developed in the USSR in 1980s. The main features of this technology were taken from the technological process of manufacturing titanium semi-finished products. Chosen way allowed successfully obtaining ingots and rods, but for the wire production usage of "titanium" technology was technologically inadvisable due to the huge number of passes and intermediate annealing during cold drawing. In this case industrial center MATEK-SMA Ltd., decided to improve this technology, developed new equipment and temperature-deformation regimes of drawing, in order to enhance production capacity and properties of the obtained wire.

Technological process

Hot forged rods or wire with a diameter of 2.5–3.0 mm, obtained on the rotary forging machine (RFM), are used as billets for drawing. Firstly, a one-pass gas-heated drawing machine "MV-3000 VM" produced by "Almaty Heavy Machine Building Plant" JSC was used to produce a wire with a diameter of 1.0–2.8 mm. For production of wire with a diameter of 0.5–1.0 mm

Published under license by Materials Research Forum LLC.

Materials Research Forum LLC
doi: http://dx.doi.org/10.21741/9781644900017-1

a similar drawing machine "MV-1000 VM" was used. Both mills were originally designed for drawing of tungsten and molybdenum. The main disadvantages of Ni-Ti wire production on these mills with applied temperature-deformation regimes were: lack of precise control of billet heating (because of gas heating) and fixed non-optimal drawing speed (due to the operational characteristics of the equipment designed to obtain wire from other alloys). Industrial center MATEK-SMA Ltd., considering all of these limitations, developed technical specifications for the manufacturing of drawing equipment specifically designed for drawing of Ni-Ti-based shape memory alloys. Based on these specifications the "194SS" and "195SS" drawing machines were manufactured. A schematic representation of the machines is shown in Fig. 1. Main technical characteristics of drawing machines are shown in table 1.

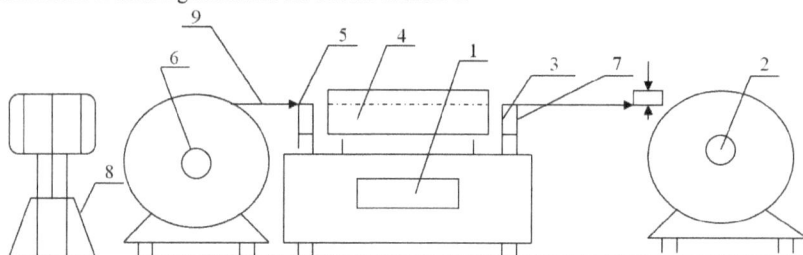

Figure 1. Schematic representation of drawing machines "194CC" and "195CC": 1) control panel; 2) traction drum; 3) die holder; 4) electric furnace; 5) oil tank; 6) winding drum; 7) die heat equipment; 8) winding system for an primary feed of a wire in the drawing machine; 9) drawing line.

Table 1. Technical characteristics/specifications of drawing mills "194CC" and "195CC".

No	Characteristics	Value	
		195SS	194SS
1	Traction drum diameter, [mm]	600	400
2	Traction drum quantity, [pcs]	1	1
3	Diameter of initial blank, [mm]	3.0	1.0
4	Diameter of a ready wire, [mm[	1.00	0.5
5	Heating equipment	electric furnace	electric furnace
6	Heating furnace length, [mm]	1600	1200
6	Drawing speed, [m/min]	0–10	0–24
7	Maximum traction drum force, [kgf]	600.0	300.0
10	Weight, [kg]	3000	2000
11	Outline dimensions, [mm]		
	Length	6270	3080
	Width	1100	900
	Height	1200	1215

Winding drums of the "194SS" and "195SS" drawing machines are equipped with the brake system to create wire tension during the drawing. Heating furnaces is made with two heating zones, two and one top covers and two thermocouples: the first is located at a distance of 300 mm from the entrance, the second – of 200 and 150 mm from the exit, respectively. The distance between the die holder and the hauling drum is sufficient for the additional installation of a water cooling bath with a length of 350 mm. Top-lifts of traction and winding drums and drums themselves are made quick-detachable and interchangeable. If wire breaks on the traction drum and the trailing end of the wire leaves the die holder automatic tripping of machines immediately activates. The developed technology of wire drawing with diameters of 0.5–2.8 mm from Ni–Ti-based shape memory alloys was optimized for this two drawing mills (Fig. 2 and Fig. 3) and may be represented by following sequence of general technological operations.

Figure 2. Drawing machine "194SS".

Figure 3. Drawing machine "195SS".

Drawing on the "195SS" drawing machine. All blanks should be sharpened before drawing, at the pointing machine: the forward end of the workpiece is sharpened for a length of 100–150 mm in two operation steps with rotation at the angle of 90 degrees.

On the next stage a blank is heated in the electric resistance furnace. The heating temperature should not exceed 600 °C (dark-cherry glow) if there is no any other requirements in technical specification of a blank. The required heating is regulated by the furnace temperature and drawing speed. Drawing speed is infinitely variable and is selected based on the particular temperature of a blank.

Drawing is performed with lubrication, usually, with aqueous colloid-graphite compound called "aquadag". If it is necessary to perform drawing without heating, greases, should be used as a lubricant, for example "Litol-24" (GOST 21150-87).

Sharpened forward end of a blank by manual tongs is set in the die, preliminary heated to 200 °C. Then this end is fixed in the filling tongs, which, in their turn, are fixed at the traction drum. At the following step a blank is placed into the furnace, the die is inserted into the die holder and a jaw clutch is simultaneously turned on. From the winding drum, a blank passes through the oil tank to the furnace, where it is dried and heated, and then goes through the die and coils on the traction drum. After the end of first drawing step, the drive of the mill is automatically switched off, the traction drum is declutched from the power shaft and two wheels are removed and replaced. Then a blank is sharpened again and drawing is carried out in the next die. The number of passes depends on the initial diameters of a blank and required diameter of a ready wire. For the one pass, the diameter of a blank reduces at 0.1 mm. The diameter of a wire is measured by a micrometer with a 0.01 mm division value after each pass. Permanent temperature control during all technological operations is also provided.

Drawing on the "194SS" drawing mill. The technological stage sequence is practically the same for "194SS" drawing machine. The main difference consists in the method of wire sharping. It is more convenient to use pickling in a special device, against the pointing machine, because of the small initial diameter of a blank. The forward 40–50 mm long end of a blank is fixed in nipping pliers and put into a cup of acid in contact with the electrode. The chemical reaction leads to the reduction of a diameter and it becomes possible to set a blank into the die. The value of reduction by the one pass for "194SS" drawing mills equals 0.05 mm. Other technological stages and parameters are directly the same both for "195SS" and "194SS" drawing machines with one.

Actual drawing schedule for wire with a diameter of 0.63 mm obtained from Ti-55.83 wt.% Ni alloy is shown in table 2. Obtained wire was supplied to the customer in accordance with technological specifications. Mechanical and functional properties of the wire are shown in table 3.

Table 2. Actual drawing schedule for wire with a diameter of 0.63 mm from rod with a diameter of 5.5 mm.

Equipment	Initial diameter, [mm]	Operation	Temperature, [°C]			Speed, [m/min]	Actual diameter, [mm]
RFM	5.5	Forging	870			1.7	4.70 ± 0.1
	4.70		-«-			--	4.30 ± 0.1
	4.30		-«-			-«-	3.80 ± 0.1
RFM	3.80		850			3.0	3.10 + 0.1
	3.10		-«-			-«-	2.60 + 0.1
	2.60		-«-			-«-	2.30 – 0.05
195SS	2.30 – 0.05/2.30	Drawing	710	620	500	2.8	2.27
	2.27/2.20		-«-	-«-	-«-	3.0	2.22
	2.22/2.10		-«-	-«-	-«-	-«-	2.11
	2.11/2.00		690	-«-	-«-	-«-	2.01
	2.01/1.90		-«-	-«-	-«-	-«-	1.90
	1.90/1.80		650	-«-	-«-	4.0	1.81
	1.81/1.70		-«-	-«-	-«-	-«-	1.71
	1.71/1.60		-«-	-«-	-«-	-«-	1.63
	1.63/1.50		-«-	-«-	-«-	-«-	1.53
	1.53/1.40		-«-	-«-	-«-	-«-	1.44
Furnace No 1	1.44/1.44	Annealing	650	-	-	30 min, air cooling	-
195SS	1.44/1.30	Drawing	650	620	500	4.0	1.33
	1.33/1.20		-«-	-«-	-«-	-«-	1.21
	1.21/1.10		580	570	500	8.0	1.12
	1.12/1.00		-«-	-«-	-«-	9.0	1.02
	1.02/0.90		550	520	400	10.0	0.91
	0.91/0.80		-«-	-«-	-«-	-«-	0.82
	0.82/0.75		-«-	-«-	-«-	-«-	0.76
194SS	0.76/0.70		520	520	400	14.0	0.70
	0.70/0.65		-«-	-«-	-«-	16.0	0.66
	0.66/0.60		-«-	-«-	-«-	17.0	0.63

Table 3. Mechanical and functional properties of the wire with a diameter of 0.63 mm.

No. of melting	After drawing		PDA 480 °C, 20 min			PDA 850 °C, 20 min + quenching in water			σ_{cr}, [MPa] (after drawing)	σ_B, [MPa] (after drawing)	δ, [%] (after drawing)
	A_s	A_f	A_f	R_s	R_f	M_s	A_s	A_f			
145	−21	−16	8	33	33	−19	−9	−6	550	1170	11

Mechanical and functional properties of the wire obtained by developed technology

Developed technological schedule of wire production on the manufactured equipment allow implementing principles of "warm forging" [5,6] technology for drawing in order to obtain wire with improved mechanical and functional properties. In this case, experimental work was carried out for the purpose of development, firstly, the technology of "warm drawing", and secondly, new technological conditions for semi-finished products with improved mechanical and functional properties. The billets for drawing, rods of TN-1 grade (Ti-54.93 wt.% Ni) 2.9 mm in diameter, were produced by rotary forging on the rotational forging machine.

The mechanical properties were determined at room temperature by the uniaxial tensile tests using universal tensile machine "INSTRON 3382" with a deformation rate of 4 mm/min. The following strain and stress parameters are determined: the critical stress of the reorientation of martensite σ_{cr}; dislocation yield stress $\sigma_{0.2}$; ultimate tensile strength σ_u; difference between dislocation yield stress and transformation yield stress $\Delta\sigma$; strain at the yield plateau ε_{pl}. The completely recoverable strain was estimated by a thermomechanical method using band samples. Obtained results are shown in table 4.

Table 4. Mechanical and functional properties of wire with a diameter of 0.7 mm.

Treatment	No. of sample	ε_{pl}, [%]	σ_{cr}, [MPa]	$\sigma_{0.2}$, [MPa]	$\Delta\sigma$, [MPa]	σ_u, [MPa]	δ, [%]	ε_r, [%]
Warm drawing	1	–	–	880	–	1358	15	No shape memory effect near room temperature
	2	–	306	882	576	1345	15,5	
Warm drawing +450 °C, 1 h	1	7.0	330	819	489	1211	35	6.0
	2	8.0	240	860	620	1219	35	
Warm drawing +450 °C, 2 h	1	9.0	121	684	563	972	45	6.5

The mechanical behavior during tensile tests at room temperature after drawing is characterized by strong deformation hardening and lack of a yielding plateau in stress-strain diagrams, which appears only after annealing. Post-deformation annealing (PDA) leads to the slight softening because of decrease of dislocation density, and appearance of shape memory effect near room temperature. The maximum ultimate tensile strength is observed after drawing (σ_B = 1358 MPa). But the elongation to fracture after this regime was only 15 %. The most attractive combination of strength and plastic properties was achieved after addition of PDA at 450 °C for 1 h (σ_B = 1211 MPa, δ = 35 %). Maximum completely recoverable strain of 6.5 % was obtained after drawing plus post-deformation annealing 450 °C, 2 h.

Summary

Developed technology of drawing, including applied equipment with particular technical characteristics and temperature-deformation regimes allow enhancing production capacity and manufacturing wire with a diameter of 0.5–2.8 mm from Ni–Ti-based shape memory alloys with high mechanical and functional properties that complied with technological conditions No 18.4270-001-16980791–2013.

Acknowledgment

The present work was carried out according to the State Task No. 007-00129-18-00.

References

[1] Q. Sun, R. Matsui, K. Takeda, E. Pieczyska, Advances in Shape Memory Materials: in Commemoration of the Retirement of Professor Hisaaki Tobushi, Springer, 2017, V.73. https://doi.org/10.1007/978-3-319-53306-3

[2] N. Resnina, V. Rubanik (Eds.), Shape Memory Alloys: Properties, Technologies, Opportunities, Trans. Tech. Publications, Praffikon, 2015.

[3] V. Brailovski, S. Prokoshkin, P. Terriault, F. Trochu (Eds.), Shape memory alloys: fundamentals, modeling and applications, Montreal, ETS Publ., 2003.

[4] K. Otsuka, X. Ren, Physical metallurgy of Ti–Ni-based shape memory alloys, Prog. mater. sci. 50 (2005) 511-678. https://doi.org/10.1016/j.pmatsci.2004.10.001

[5] S. Prokoshkin, I. Khmelevskaya, V. Andreev, R. Karelin, V. Komarov, A. Kazakbiev, Manufacturing of Long-Length Rods of Ultrafine-Grained Ti-Ni Shape Memory Alloys, In Mater. Sci. F. 918 (2018) 71-76. https://doi.org/10.4028/www.scientific.net/MSF.918.71

[6] V. Andreev, V. Yusupov, M. Perkas, V. Prosvirnin, A. Shelest, S. Prokoshkin, I. Khmelevskaya, A. Korotitskii, S. Bondareva, R. Karelin, Mechanical and functional properties of commercial alloy TN-1 semiproducts fabricated by warm rotary forging and ECAP, Russ. Met. (Metally). 10 (2017) 890-894. https://doi.org/10.1134/S0036029517100020

Shape Memory Alloys – SMA 2018
Materials Research Proceedings **9** (2018) 9-13

Materials Research Forum LLC
doi: http://dx.doi.org/10.21741/9781644900017-2

Production of Biocompatible TiNi-based Porous Materials with Terraced Surface of Pore Walls

Sergey G. Anikeev[1,a*], Nadezhda V. Artyukhova[1,b], Valentina N. Khodorenko[1,c], Alexander S. Garin[1,d], Victor E. Gunther[1,e]

[1]National Research Tomsk State University, 36 Lenin Ave, Tomsk 634050, Russia

[a]Anikeev_Sergey@mail.ru, [b]Artyukhova_nad@mail.ru, [c]Hodor_val@mail.ru, [d]Stik-020@mail.ru, [e]Nii_mm@sibmail.com

*corresponding author

Keywords: TiNi-based Alloy, Sintering, Powder Metallurgy, Structure, Martensite, Biocompatible Implants

Abstract. The structural features of the sintered TiNi-based porous materials with terraced morphology of the pore wall surface have been studied. A porous alloy with the TiNi intermetallic compound in a two-phase state was produced by diffusion liquid-phase sintering of TiNi powder. In the samples obtained, the regions of melt formation exhibit a terraced relief of the pore wall surface. The terraces are located in the regions free of secondary-phase particles and propagate along the curved surface of the pore walls within one grain. Hexagonal areas are observed in some regions of the pore wall surface. The terraced relief is formed in the regions where the martensite crystals B19' emerge on the surface of the pore walls.

1. Introduction

TiNi-based alloys refer to materials with a wide range of structural and functional properties. The combination of high parameters of biochemical compatibility due to oxide layers on the surface of TiNi alloys and biomechanical compatibility due to the hysteretic nature of the shape change under alternating stress makes this material suitable for production of implantable structures [1].

Powder metallurgy methods employed to produce TiNi-based alloys solve one of the main problems of medical materials science focused on their structural correspondence to the bulk tissues in the human body. Self-propagating high-temperature synthesis (SHS) and sintering are used to produce biocompatible porous TiNi-based materials [2–4]. Porous TiNi alloys imparted with the properties of shape memory and superelasticity in combination with the developed permeable structure of the pore space can be used to produce implants with high functional properties to solve a wide range of medical problems [5]. In contrast to SHS, porous TiNi-based materials are produced by diffusion liquid-phase sintering to create small-sized implants with lower porosity and reduced pore size interval.

Studies are under way to improve the structural parameters of porous TiNi-based materials in order to optimize the integration of porous permeable structures in the implant–tissue system. One of the methods to improve integration is to form a developed rough surface of the pore walls, which increases the adhesive properties of the material and helps to attach more cell populations. A high amount of the cell mass at the early stages of its development in the pore space of implants facilitates differentiation and growth of the formed tissues. The aim of the study was to develop the method to increase the parameters of the pore wall roughness in porous TiNi-based materials produced by diffusion liquid-phase sintering.

Published under license by Materials Research Forum LLC.

Shape Memory Alloys – SMA 2018   Materials Research Forum LLC
Materials Research Proceedings **9** (2018) 9-13   doi: http://dx.doi.org/10.21741/9781644900017-2

2. Experimental procedures

Porous TiNi-based materials were produced by single diffusion liquid-phase sintering of TiNi powders of grade PV–N55T45 (PJSC Tulachermet, Tula). Single-step sintering in quartz tubes was used to produce samples at temperatures in the range of 1220–1270 °C and sintering time of 15 min. During the production of porous materials, temperature-time conditions that favor the formation of a moderate amount of melt are created due to melting of the Ti_2Ni phase and partial dissolution of the TiNi phase. To keep the powder inside the quartz tubes, we used holders made from porous SHS TiNi alloy. Sintering was carried out at 6.65×10^{-4} Pa and an average heating rate of 10 °C/min. The initial porosity of the fill before sintering was 65–70 % with a diameter of quartz tubes of 12–14 mm.

The properties of the initial powder significantly affect the structural parameters and physico-mechanical properties of the materials produced through sintering. Therefore, structural characteristics and morphological features of the initial TiNi powder of grade PV–N55T45 were additionally investigated in order to interpret the results obtained. To study the macro- and microstructure of powder samples and porous TiNi alloy, metallographic sections were prepared. The samples were placed in an epoxy resin based polymer, and after polymerization they were grinded using sandpaper within the grain size interval (P493–P5000) with abundant water cooling. After each grinding step, the surface quality was monitored using the Axiovert-40 MAT optical microscope. To reveal the structural features of the test samples, the surface of metallographic sections was treated in aqueous solution of nitric and hydrofluoric acids. The etching regime was chosen experimentally, excluding excessive etching of the sample microstructure.

A quantitative description of the pore structure (specific surface area, average pore size) was made, and the pore size distribution was constructed by the secant method for samples produced at temperatures of 1250, 1260 and 1270 °C, since other samples were destroyed during grinding. The porosity was determined by weighing. The analysis of data and the construction of histograms were carried out using Origin, Statistica and GraphPad Prism software. The surface structure of the powder particles and pore walls, and the microstructure of the metal scaffold of the produced samples were investigated by optical and scanning electron microscopy (SEM) at accelerating voltages of 20–30 kV using a Quanta 200 3D Dual Beam system. For SEM studies of powder samples, pre-magnetron sputtering of 300–360 Å gold films was performed using the SPI-Module Sputter Coater equipment. The composition of the phases was determined using an EDAX ECON IV energy dispersive spectrometer (EDS). X-ray diffraction (XRD) analysis was carried out using Shimadzu XRD 6000 and ARL X'TRA X-ray diffractometers equipped with a semiconductor detector with Cu radiation.

3. Results and discussion

Sintering of TiNi powders of grade PV–N55T45 for the production of porous permeable materials can be formally referred to one-component liquid-phase sintering. However, in practice it is necessary to take into account the complex phase composition of the TiNi-based powders produced by hydride-calcium reduction. X-ray diffraction analysis was used to establish a set of phase states in the composition of TiNi powder, which are predictable for the investigated Ti–Ni system. The TiNi intermetallic compound exhibits two phases – B2-austenite and B19′-martensite. The Ti-rich phase Ti_2Ni and Ni-rich phases $TiNi_3$ and Ti_3Ni_4 were detected.

TiNi powder has a dual spongy and compact particle morphology (Fig 1). The spongy structure of powder particles is formed during the reduction reaction of titanium dioxide with calcium [6]. These particles are characteristic of this type of powders. Spongy particles consist of the TiNi phase grains of up to 30 μm in size with rectified boundaries and triple junctions that form a 60° angle. Compact particles consist of coarsely rounded TiNi grains with a massive interlayer of Ti_2Ni compound. In these particles, the ensemble of single TiNi grains can contain several large grains from 40 to 130 μm in size. The energy-dispersive microanalysis revealed that

Shape Memory Alloys – SMA 2018   Materials Research Forum LLC
Materials Research Proceedings **9** (2018) 9-13   doi: http://dx.doi.org/10.21741/9781644900017-2

TiNi grains in compact particles are surrounded by the Ti_2Ni phase formed through a peritectic reaction.

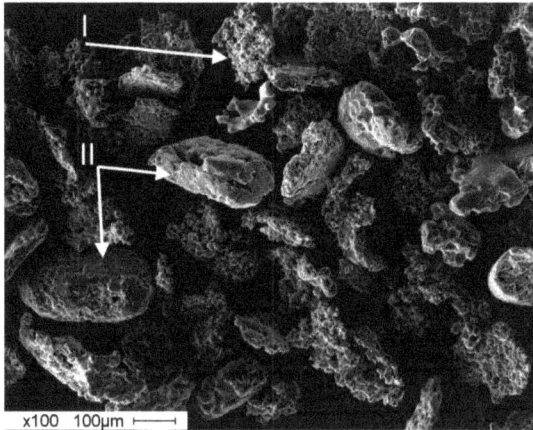

Figure 1. Dual morphology of TiNi-based powder particles: spongy (I) and compact (II).

The structural-morphological variants of the martensite phase B19' in the initial TiNi powder are of great interest [7]. A martensite structure composed of orthogonal twin plates located in different orientations within one massive rounded grain can be observed in compact particles. Spongy particles exhibit a different type of martensitic structure composed of multiple crystals of B19' martensite with a pyramidal lath morphology. They fill the entire grain volume in different orientations. These crystals minimize elastic stresses in the grain volume due to the variety of orientations and sizes of martensite crystals [8]. The structure of martensite of this type of morphology indicates the presence of internal stress fields in the powder particle volume. In addition, the crystals of the martensite phase emerge at the periphery, which indicates the presence of stress fields on the particle surface [8].

According to the diagram illustrating the state of the Ti-Ni system, the first batches of melt appear at a temperature of 955 °C during the Ti_2Ni phase melting [9]. During sintering, the contacts between the powder particles are formed as a result of surface diffusion of atoms and local wetting of the powder contacts. Due to a significant fraction of the Ti_2Ni phase, its volume is sufficient to form the initial interparticle contacts at given temperatures. Further increase in sintering temperatures close to 1240 °C leads to the formation of a new source of TiNi phase based melt. An increase in the sintering temperature in the range of 1240–1270 °C causes significant macro- and microstructural changes, which are manifested such as a decreased index of porosity, a reduced specific surface area, and increased average size of macropores and pore size interval.

All samples of the porous TiNi-based alloy produced by diffusion liquid-phase sintering exhibit the presence of the developed terraced surface of macropore walls (Fig. 2). The terraces are located in the areas free of secondary phase particles, and they propagate along the curved surface of the pore walls within one grain. In the transition from one grain to another, the direction of terraces may change or remain the same, depending on the grain orientation. In some places, a regular hexagonally-shaped area is detected. It is surrounded by curved terraces that extend along the concave or convex surface of the pore walls.

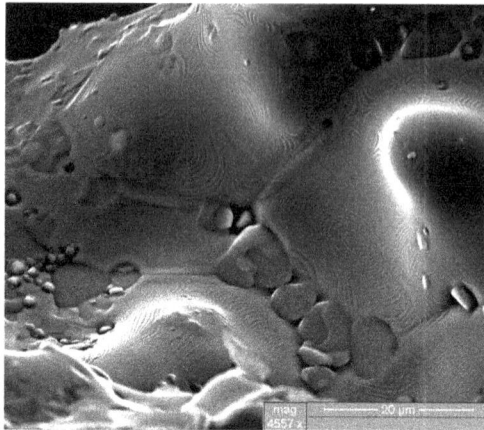

Figure 2. Terraced surface of macropore walls of the porous TiNi-based alloy produced at the temperature 1260 °C by diffusion liquid-phase sintering.

According to the *Terrace* Ledge Kink *model* (TLK), the relief is formed by attaching adatoms to the kink surfaces found on the surface of crystalline bodies [10,11]. This relief is formed due to the surface diffusion of adatoms, volume diffusion of atoms and their interaction with the substrate defects (dislocations, twins, kinks, ledges, grain boundaries, interfaces, secondary inclusions) during melt crystallization. In this case, martensite plates B19' are structural defects.

Summary

Thus, it was found that the initial TiNi-based powder has a complex structural-phase composition. During sintering, the power is homogenized, which leads to an increase in the fraction of the major TiNi phase. In the samples produced, all areas of melt formation exhibit a terraced relief on the surface of the pore walls, which is formed in the regions where the martensite B19' crystals emerge on the surface of the pore walls. The formation of a terraced relief on the surface of the pore walls in the micron range increases the pore wall roughness, which positively affects the adhesive properties of the alloy in the living organism.

Acknowledgements

This investigation has been funded by the Russian Science Foundation (Project No. 17-79-10123).

References

[1] V.E. Gunther, V.N. Khodorenko et al., Shape Memory Medical Materials. Shape Memory Medical Materials and Implants, first ed., NPTS MITS, Tomsk, 2011.

[2] V.N. Khodorenko, S.G. Anikeev, O.V. Kokorev et al., An Investigation of Porous Structure of TiNi-Based SHS-Materials Produced at Different Initial Synthesis Temperatures, Russ. Phys. J. 60 (2018) 1758-1767. https://doi.org/10.1007/s11182-018-1279-8

[3] S. G. Anikeev, V. N. Hodorenko, T. L. Chekalkin et al.,. Fabrication and study of double sintered TiNi-based porous alloys, Smart Mater. Struct. 26 (2017) 057001. https://doi.org/10.1088/1361-665X/aa681a

Materials Research Forum LLC
doi: http://dx.doi.org/10.21741/9781644900017-2

[4] S.G. Anikeev, V.N. Khodorenko, O.V. Kokorev et al., Study of Structural Features of Porous TINI-Based Materials Produced by SHS and Sintering, Adv. Mat. Res. 1085 (2015) 430-435. https://doi.org/10.4028/www.scientific.net/AMR.1085.430

[5] V. Gunther et al., Delay Law and New Class Of Materials and Implants in Medicine, MA: STT, Northampton, 2000.

[6] S.S. Kiparisov, Powder Metallurgy, Metallurgiya, Moscow, 1980.

[7] S.G. Anikeev, A.S. Garin, N.V. Artyukhova et al., Structural and Morphological Features of TiNi-Based Powder Manufactured by the Method of Hybrid-Calcium Reduction, Russ. Phys. J. 61 (2018) 749-756. https://doi.org/10.1007/s11182-018-1456-9

[8] C.M. Wayman, Proc. Int. Symp. SMA–86 (1986) 59-72. https://doi.org/10.1557/PROC-86-59

[9] K. Otsuka, X Ren, Physical metallurgy of Ti–Nibased shape memory alloys Prog. Mater. Sci. 50 (2005) 511–678. https://doi.org/10.1016/j.pmatsci.2004.10.001

[10] O.G. Kozlova. Growth and morphology of crystals, MGU, Moscow, 1980.

Shape Memory Alloys – SMA 2018
Materials Research Proceedings 9 (2018) 14-18

Materials Research Forum LLC
doi: http://dx.doi.org/10.21741/9781644900017-3

Investigation of Intermetallic Alloys Based on Ni-Mn with Controlled Shape Memory Effect

Elena S. Belosludtseva[1,a*], Vladimir G. Pushin[1,2,b], Elena B. Marchenkova[1,c],
Alexey E. Svirid[1,d], Artemiy V. Pushin[1,e]

[1]M.N. Miheev Institute of Metal Physics of Ural Branch of Russian Academy of Sciences,
18 S. Kovalevskaya Str., Ekaterinburg 620137, Russia

[2]Ural Federal University named after the first President of Russia B.N.Yeltsin,
19 Mira Str., Ekaterinburg 620002, Russia

[a]ebelosludceva@mail.ru, [b]pushin@imp.uran.ru, [c]em1104@mail.ru, [d]svirid2491@rambler.ru,
[e]avpushin@rambler.ru

*corresponding author

Keywords: Shape Memory Alloys, Thermoelastic Martensitic Transformation, Magnetoelastic Martensitic Transformation, Intermetallic Alloys, Structure

Abstract. Systematic studies of structural and magnetic phase transformations and properties of alloys of four basic quasibinary systems based on NiMn-NiTi, NiMn-NiAl, NiMn-NiGa and Ni_2MnGa-Ni_3Ga have been presented. Common phase diagrams of the existence of structural and magnetic phase transitions are constructed. It has been established that alloying with a third component (Ti, Al, Ga or Mn) lowers the critical temperatures of thermoelastic martensitic transformations (TMT) related shape memory effects (SME) and pseudoelasticity (PEE) with respect to the base intermetallides.

Introduction

$Ni_{50}Mn_{50}$ and $Ni_{49}Mn_{51}$ alloys undergo martensitic transformation at high temperatures, which is of particular interest for studying the structure and properties of these alloys in the transformation temperature range. In [1,2], we comprehensively investigated the structure and the physical properties of these alloys, revealed a thermoelastic mechanism of the martensitic transformation, and determined the critical temperatures of the thermoelastic martensitic transformation (TMT) in them ($M_s = 970$, $M_f = 920$, $A_s = 970$, $A_f = 1020$, $M_s = 940$, $M_f = 930$, $A_s = 990$, $A_f = 1000$ K). The high-temperature B2 $\leftrightarrow$ L1$_0$ phase transformation is known to occur in many binary and multicomponent intermetallic alloys based on nickel and titanium, such as Ni–Mn, Ni–Al, Ni–Mn–Al, Ni–Al–Co, Ti–Rh, Ti–Ir, Ti–Rh–Ni, and Ti–Ir–Ni [1-14]. There are good foundations to believe that this transformation in alloys based on such intermetallic compounds and in other B2 alloys (nonferrous titanium nickelide, copper-based alloys, ferromagnetic Heusler alloys, alloys based on alloyed manganese nickelide) also has signs of TMT, and this fact should cause the shape memory effect in them [4,5]. In this transformation, the volume effect is 1.6 % [7-9]. Nevertheless, the boundaries of phase transformations and the crystal structure types of austenite and martensite phases have not been exactly determined even for ternary ferromagnetic Ni–Mn–Ga alloys, the single crystals of which exhibit giant, up to 10%, magnetically induced reversible strains [10]. In the binary alloys of Ni-Al and ternary Ni-Mn-Al and Ni-Mn-Ti mechanical properties were measured for compression and SMAs were found that were small in magnitude of reversible deformation in Ni-Al alloys (less than 0.3 %) and significant in its (up to 5 %) for Ni-Mn alloys [10,11,13], with compression deformation up to fracture up to 10–15% [11,12].

Published under license by Materials Research Forum LLC.

Shape Memory Alloys – SMA 2018
Materials Research Proceedings 9 (2018) 14-18

Materials Research Forum LLC
doi: http://dx.doi.org/10.21741/9781644900017-3

The purpose of this work is to study the structure and the phase transformations in a number of quasi-binary Ni-Mn alloys and to determine the alloying effect on the structure, the phase composition, TMT, and the critical TMT points during the forward (M_s, Mf) and reverse (A_s, A_f) martensitic transformation in these alloys.

Results

According to X-ray diffractometry data, the phase composition, the structural types of austenite and martensite, and the parameters [6] of their structure are determined (Fig. 1). Phase decomposition of alloys was found in Ni-Mn-Ti alloys with containing Ti more than 15 at.%. The second phase is based on Ni_3Ti with manganese dopping.

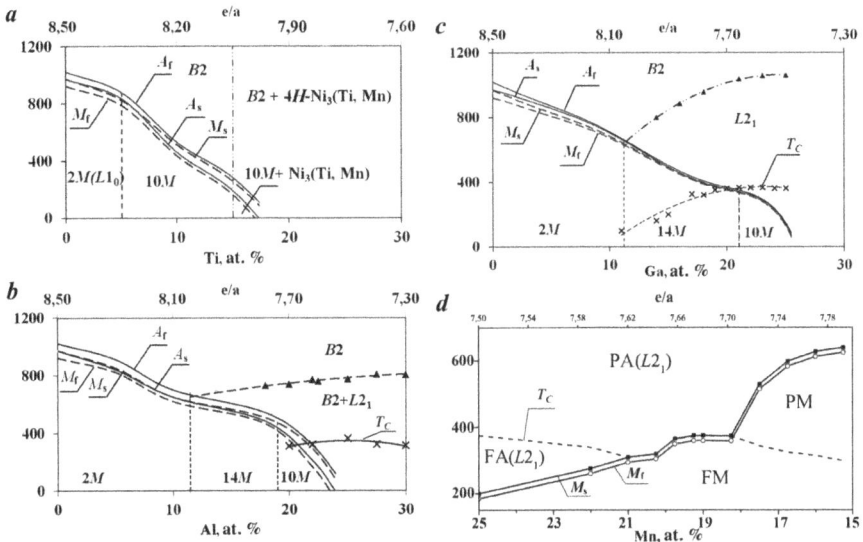

Figure 1. Common phase diagrams of investigated systems of alloys: a – Ni-Mn-Ti, b – Ni-Mn-Al, c – Ni-Mn-Ga, d – Ni-Ga-Mn.

Transmission (TEM) and scanning (SEM) electron microscopy methods were used to obtain typical images of 2M and 14M martensite microstructures and high atomic resolution images (Fig. 2). Microscopic micrographs of martensite showed twin-type reflexes or extra reflexes located at a distance of 1/5 or 1/7 between the main Bragg reflections that corresponded to the 10M or 14M phase, respectively. A joint trace analysis of TEM images and microelectronograms made it possible to establish orientational phase relationships. It is shown that martensite consists of packets of pairwise twinned parallel plates with a habit plane close to the {011} B2 or L21-metastable austenite and thin secondary internal twins in one of 24 equivalent "soft" twinning shift systems {011} <01$\bar{1}$> $_{B2 / L21}$ (Fig. 2, 3). On the basis of the obtained microstructural data, the crystal structure mechanism of the TMT is proposed by a regular planar homogeneous atomic shift in the direction of the {011}<01$\bar{1}$> $_{BCT}$ or {111}<11$\bar{2}$> $_{FCT}$ type. In this case, in contrast to the Baine scheme, the orientation relations B2 (L2$_1$) and phases 2M, 10M, 14M, observed in the experiment, are realized (Fig. 3). The existence of a hierarchy of packets of twinned crystals in these alloys is due to the action of the multiple-nucleus mechanism of TMT and subsequent

Shape Memory Alloys – SMA 2018   Materials Research Forum LLC
Materials Research Proceedings 9 (2018) 14-18   doi: http://dx.doi.org/10.21741/9781644900017-3

adaptive twinning, which can develop upon cooling and in the martensitic phase. The main reason for the formation of a fairly well-organized hierarchy of coherent twin crystals in the studied low-modulus alloys is the anisotropic elastic stresses accumulated at such TMT. The thermoelastic mechanism of transformation causes a low density of dislocations in martensite phases 2M, 10M and 14M, repeatability in the thermocycling of their micromorphology and, most importantly, the presence of SME and PEE in them.

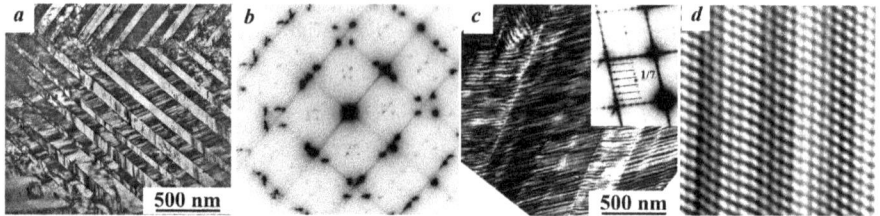

Figure 2. Microstructure (a, c), electron diffractions (b, insert on c) and direct atomic resolution images (d) of alloys $Ni_{50}Mn_{50}$ (a, b) and $Ni_{50}Mn_{32}Al_{18}$ (c, d) with 2M- and 14M-martensites, respectively.

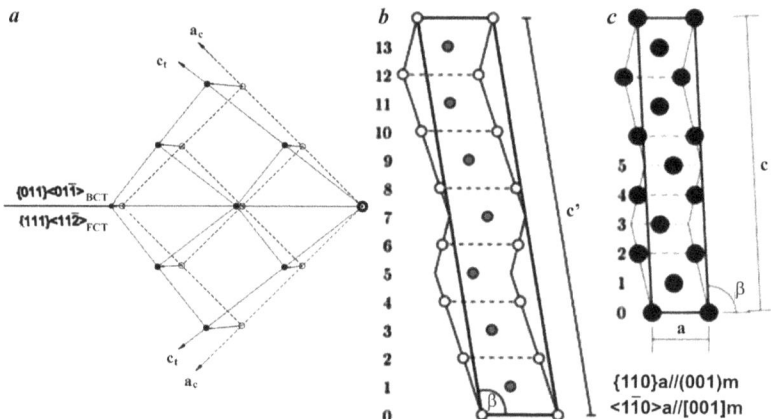

Figure 3. Schematic representation of the lattice of 2M (a), 14M (b), and 10M (c) crystal lattices [14].

For alloys of all systems, the dependence of the structural type of the martensitic phases on the electron concentration was observed. If in quasi-binary alloys with Al and Ga e/a is 8.50–8.10, then martensite is ordered as 2M if $8.10 > e/a > 7.70$, then the structure of martensite is of the 14M type, with $7.70 > e/a > 7.60$ – to 10M. When e/a more then 7.60 martensitic transformation does not happen. For Ti alloys, other concentration boundaries have been found: $8.50 > e/a > 8.35 – 2M$ ($L1_0$) martensite, $8.35 > e/a > 8.05 – 10M$, $e/a \leq 8.05 – B2$-austenite.

Shape Memory Alloys – SMA 2018   Materials Research Forum LLC
Materials Research Proceedings **9** (2018) 14-18   doi: http://dx.doi.org/10.21741/9781644900017-3

Summary

Since the temperatures of magneto-, mechano- and thermally controlled martensitic transformations in the studied alloys vary in a very wide range, they can be used in various fields of technology, selecting the chemical composition depending on the required performance characteristics.

Acknowledgements

The work was carried out within the framework of the state task of Federal Agency for Scientific Organizations (FASO Russia) (cipher "Structure", No. AAAA-A18-118020190116-6) and with partial support of the RFBR (project No. 18-32-00529 mol_a).

References

[1] K. Ootsuka, K. Simidzu, Yu. Sudzuki, et al., Shape Memory Alloy, Ed. by Kh. Funakubo, Kyoto, 1984, Gordon and Breach Science, New York, 1987.

[2] K. Adachi and C.M. Wayman, Electron microscopic study of Θ-phase martensite in Ni-Mn alloys Metall. Trans. A 16 (1985) 1581-1597. https://doi.org/10.1007/BF02663014

[3] K. Adachi and C.M. Wayman, Transformation behavior of nearly stoichiometric Ni-Mn alloys Metall. Trans. A 16 (1985) 1567-1579. https://doi.org/10.1007/BF02663013

[4] A.G. Popov, E. V. Belozerov, V. V. Sagaradze N.L. Pecherkina, I.G. Kabanova, V.S. Gaviko, V.I. Khrabrov, Martensitic transformations and magnetic-field-induced strains in $Ni_{50}Mn_{50-x}Ga_x$ alloys, Phys. Met. Metallog. 102 (2006) 140-148. https://doi.org/10.1134/S0031918X06080047

[5] N.I. Kourov, V.G. Pushin, A.V. Korolev, V.V. Marchenkov, E.B. Marchenkova, V.A. Kazantsev, H.W. Weber, Effect of Severe Plastic Deformation by Torsion on Structure and Properties of $Ni_{54}Mn_{21}Ga_{25}$ and $Ni_{54}Mn_{20}Fe_1Ga_{25}$, Sol. Stat. Phenomena 168-169 (2011) 553-556. https://doi.org/10.4028/www.scientific.net/SSP.168-169.553

[6] E.S. Belosludtseva, N.N. Kuranova, E.B. Marchenkova, V.G. Pushin, Features of Thermoelastic Martensitic Transformations, Structure and Properties in ternary B2-alloys based on NiMn – NiTi, NiMn – NiAl, NiMn – NiGa, Ni_2MnGa – Ni_3Ga quasi-binary system, Materials Today: Proceedings. 4 (2017) 4717-4721. https://doi.org/10.1016/j.matpr.2017.04.058

[7] P.L. Potapov, N.A. Polyakova, V.A. Udovenko, E.L. Svistunova, The martensitic structure and shape-memory effect in NiMn alloyed by Ti and Al, Z. Metallkd. 87 (1996) 33-39

[8] V.G. Pushin, E.S. Belosludtseva, V.A. Kazantsev, N.I. Kourov, Transformation and Fine Structure of Intermetallic Compound $Ni_{50}Mn_{50}$, Inorganic Materials: Applied Research, 4 (2013) 340-347. https://doi.org/10.1134/S2075113313040084

[9] V.G. Pushin, N.N. Kuranova, E.B. Marchenkova, E.S. Belosludtseva, V.A. Kazantsev, and N.I. Kourov, High-Temperature shape memory effect and the B2-L1$_0$ thermoelastic martensitic transformation in Ni-Mn intermetallics, Tech. Phys. 58 (2013) 878-887. https://doi.org/10.1134/S1063784213060236

[10] J. Pons, E. Cesari, C. Segu´i, F. Masdeu, R. Santamarta, Ferromagnetic shape memory alloys: Alternatives to Ni–Mn–Ga, Mat. Sci. Eng. A 481-482 (2008) 57-65. https://doi.org/10.1016/j.msea.2007.02.152

[11] T. Krenke, M. Acet, E.F. Wassermann, X. Moya, L. Mañosa, A. Planes, Martensitic transitions and the nature of ferromagnetism in the austenitic and martensitic states of Ni-Mn-Sn

alloys, Phys. Rev. B 72 (2005) 014412-1-014412-9.
https://doi.org/10.1103/PhysRevB.72.014412

[12] Y.M Jin, Yu U. Wang, Y. Ren, Theory and experimental evidence of phonon domains and their roles in pre-martensitic phenomena, Nature Partner Journals: Computational Materials 1, Article number: 15002 (2015) - http://www.nature.com/articles/npjcompumats20152.

[13] K.K. Lee, P.L. Potapov, S.Y. Song, M.C. Shin, Shape memory effect in NiAl and NiMn based alloys, Scripta Mater. 36 (1997) 207-212. https://doi.org/10.1016/S1359-6462(96)00363-6

[14] T. Büsgen, J. Feydt, R. Hassdorf, S. Thienhaus, M. Moske, M. Boese, A. Zayak, P. Entel, Ab initio calculations of structure and lattice dynamics in Ni−Mn−Al shape memory alloys, Phys. Rev. B. 70 (2004) 014111-1-014111-8. https://doi.org/10.1103/PhysRevB.70.014111

Shape Memory Alloys – SMA 2018
Materials Research Proceedings **9** (2018) 19-23

Materials Research Forum LLC
doi: http://dx.doi.org/10.21741/9781644900017-4

Mechanical and Functional Properties of Ti$_{48.6}$Ni$_{49.6}$Co$_{1.8}$ Shape Memory Alloy

Roman D. Karelin[1,2,a*], Vladimir A. Andreev[2,4,b], Irina Yu. Khmelevskaya[1,c],
Sergey D. Prokoshkin[1,d], Natalia N. Resnina[3,e], Victor S. Komarov[1,f],
Sofya A. Bondareva[1,g], Vladimir S. Yusupov[2,h]

[1]National University of Science and Technology "MISiS", 4 Leninskiy Prospect, Moscow
119049, Russia

[2]Baikov Institute of Metallurgy and Materials Science, RAS, 49 Leninskiy Prospect, Moscow,
119991, Russia

[3]Saint-Petersburg State University, 7/9 Universitetskaya Naberezhnaya, Saint-Petersburg
199034, Russia

[4]MATEK-SMA Ltd., 2a-137 Karier Str., Moscow 117449, Russia

[a]rdkarelin@gmail.com, [b]andreev.icmateks@gmail.com, [c]khmel@tmo.misis.ru,
[d]prokoshkin@tmo.misis.ru, [e]resnat@mail.ru, [f]komarov@misis.ru, [h]yusupov@aport2000.ru,
[g]sonya60@rambler.ru

*corresponding author

Keywords: Shape Memory Alloys, Titanium Nikelide, Thermomechanical Treatment, Rotary Forging, Functional Properties, Superelasticity

Abstract. Mechanical and functional properties of Ti$_{48.6}$Ni$_{49.6}$Co$_{1.8}$ shape memory alloy rods with diameters of 3.5 and 4.5 mm were studied after rotary forging (RF). The structure was analyzed using optical microscopy. Mechanical properties were determined by uniaxial tensile and Rockwell hardness tests. Temperature range of martensitic transformations was defined by differential scanning calorimetry. Superelasticity effect was studied by a thermomechanical method using bending tests. The obtained results showed that RF at 450 °C and RF + annealing at 450 °C, 1 hour allow manufacturing 3 m-long Ti-Ni-Co rods with diameters of 3.5 and 4.5 mm with improved mechanical properties: (RF: σ_B = 1180 MPa, $\sigma_{0.2}$ = 625 MPa, δ = 15 % and σ_B = 1205 MPa, $\sigma_{0.2}$ = 622 MPa, δ = 16 %, respectively; RF+ annealing: σ_B = 1292 MPa, $\sigma_{0.2}$ = 651 MPa, δ = 19 % and σ_B = 1228 MPa, $\sigma_{0.2}$ = 575 MPa, δ = 21 %, respectively). The temperature range of superelastic behavior was determined as –30 to 20 °C with the value of completely maximum superelastic strain – 5.0 %.

Introduction

Ti-Ni-based shape memory and superelastic alloys (SMA) are widely used in different fields of engineering and medicine as a functional material for production of various shape-memory devices [1]. The most frequently used SMA is binary Ti-Ni alloy. It is well-known that by addition of ternary element it is possible to vary mechanical and functional properties in order to obtain required operational characteristics. In previous studies it was shown that addition of Co as a substitute for Ti for more than 0.5 % leads to the decrease in the temperature range of martensitic transformation less then temperature of liquid nitrogen. Addition of Co substitute for Ni, in their turn, leads to the increase in mechanical properties and appearance of superelasticity effect near and below room temperature [2-5]. Another effect of Co addition consists in the decrease of deformability and complicates the possibility of industrial production and application of the alloy. One of the most frequently used processes for obtaining round and long-length rods of SMA is a hot rotary forging (RF). It follows from the previous experience that for formation

Published under license by Materials Research Forum LLC.

of ultrafine-grained (submicro- and nanocrystalline) structure and considerable improvement of mechanical and functional properties, the deformation temperature should be 450 °C or less [6-8]. So that, it was necessary to investigate the possibility of production long-length Ti-Ni-Co rods with superelastic behavior in wide temperature range and high mechanical properties, by rotary forging at relatively low temperatures. In the present paper, mechanical and functional properties of Ti-Ni-Co rods produced by rotary forging at 450 °C are described and discussed.

Experimental

In the present work, $Ti_{48.6}Ni_{49.6}Co_{1.8}$ was studied. Initial samples were obtained by vacuum induction melting in industrial center "MATEK-SMA" Ltd. The billets for rotary forging in a form of rods 20 mm in diameter were produced by screw rolling at $850 – 950$ °C with reduction of $7 – 20$ % per pass and interpass heating (hot-rolled state).

Hot-rolled rods were rotary forged from a diameter of 20 to 4.5 and 3.5 mm at 450 °C with particular strains from 1 to 5 %. A post-deformation annealing at the deformation temperature for 1 and 2 hours was performed after all regimes of thermomechanical treatment. A post-deformation annealing at 850 °C, 10 and 30 min after RF was served as a reference treatment (RT).

The structure was studied at room temperature using "UNION" optical microscope. Characteristic temperatures of martensitic transformations were studied using the "Mettler Toledo" calorimeter. The Rockwell hardness measurements were carried out at room temperature using a "TR 5008" tester under a preload of 98 N (10 kg) and load of 588 N (60 kg). The mechanical properties were determined at room temperature by the uniaxial tensile tests using universal tensile machine "INSTRON 3382" with a deformation rate of 4 mm/min. The maximum superelastic strain was estimated by a thermomechanical method using band samples.

Results and Discussion

Differential scanning calorimetry. The results of the differential scanning calorimetry showed that in initial state, after vacuum induction melting, single-stage forward and reverse martensitic transformations occur, M_s, M_f and A_s locate below -100 °C. RF at 450 °C provides appearance of B2→R transformation and increases temperatures of martensitic B2↔B19′ transformation. Post-deformation annealing for 1 and 2 hours leads to a slight decrease of these temperatures (Table 1). After post-deformation annealing at 850 °C for 10 and 30 (RT) minutes, the picture is practically the same as after melting.

Optical microscopy. Microstructure of samples can be studied by optical microscopy only after RT, because of the formation of ultrafine-grained structure after warm RF and significant refinement of structural elements. In the microstructure after RT, presumably Ti_2Ni-phase type particles can be observed.

Mechanical and functional properties. The measurements of the Rockwell hardness (Fig. 2), show higher hardness value after the RT both for samples with 3.5 and 4.5 mm in comparison with the RT (64 HRA and 61 HRA respectively against 52 HRA). Post-deformation annealing at 450 °C for 1 h leads to the partial softening of internal stresses and decrease of the hardness value (55 HRA and 56 HRA, respectively).

The mechanical behavior during tensile tests at room temperature after RF is characterized by strong deformation hardening and small area of yielding plateau in stress-strain diagrams, which becomes longer after annealing at 450 °C for 1 h. Also, the strength characteristics for diameters of 3.5 and 4.5 mm are higher after post-deformation annealing (Table 2).

Shape Memory Alloys – SMA 2018   Materials Research Forum LLC
Materials Research Proceedings **9** (2018) 19-23   doi: http://dx.doi.org/10.21741/9781644900017-4

Table 1. Results of differential scanning calorimetry.

Treatment		R_s, [°C]	R_f, [°C]	M_s, [°C]	M_f, [°C]	R'_s, [°C]	R'_f, [°C]	A_s, [°C]	A_f, [°C]
Cast ingot		–	–	–106	<–110	–	–	–103	–57
⌀3.5	RF	29	–32	–60	<–110	–33	–	15	32
	RF+ 450 °C, 1 h	13	–14	–80	<–110	–31	–	0	19
	RF+ 450 °C, 2 h	15	–7	–75	<–110	–28	–	5	23
	RF+ 850 °C, 0.1 h	–	–	–97	<–110	–	–	–78	–46
	RT	–	–	–	–	–	–	–90	–17
⌀4.5	RF	38	–33	–47	<–110	–28	–	25	43
	RF+ 450 °C, 1 h	17	–13	–71	<–110	–33	–	4	20
	RF+ 450 °C, 2 h	21	–3	–61	<–110	–25	–	10	27
	RF+ 850 °C, 0.1 h	–	–	–91	<–110	–	–	–80	–40
	RT	–	–	–	–	–	–	–80	6

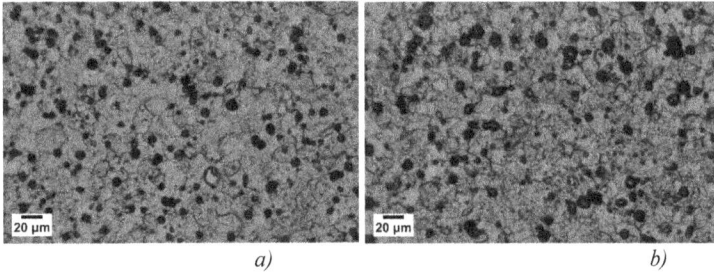

a)   b)

Figure 1. Microstructure of samples with a diameter of 3.5 (a) and 4.5 (b) mm after RT.

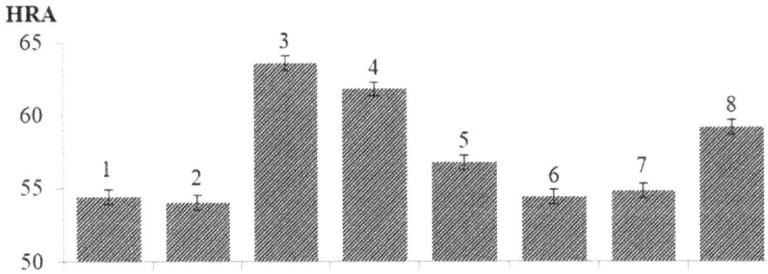

Figure 2. Rockwell hardness test results: 1 – ⌀3.5 + 850 °C, 1/2 h; 2 – ⌀4.5 + 850 °C, 1/2 h; 3 – ⌀3.5; 4 – ⌀4.5; 5 – ⌀3.5 + 450 °C, 1 h; 6 – ⌀4.5 + 450 °C, 1 h; 7 – ⌀3.5 + 450 °C, 2 h; 8 – ⌀4.5 + 450 °C, 2 h.

Table 2. Tensile tests result.

Treatment	$\sigma_{0.2}$, [MPa]	σ_{B}, [MPa]	δ, [%]
RF$_{450}$ ∅3.5	625	1180	15
RF$_{450}$ ∅3.5+ 450 °C, 1 h	651	1292	19
RF$_{450}$ ∅4.5	622	1205	16
RF$_{450}$ ∅4.5+ 450 °C, 1 h	575	1228	21

Superelasticity Effect. After rotary forging the maximum value of superelastic strain was 5.0 % for samples with a diameter of 3.5 mm and 6.0 % – of 4.5 mm. Value of superelastic shape recovery rate after 5 cycles of bending for 6.0 % at room temperature was 77 and 85 % for samples with a diameter of 3.5 and 4.5 mm, respectively. Post-deformation annealing at 450 °C for 1h leads to the increase in the maximum value of superelastic strain to 6.0 and 6.5 % and superelastic shape recovery rate to 85 and 87 % for samples with a diameter of 3.5 and 4.5 mm respectively. Superelastic behavior of strain in dependence of the temperature range from –50 to 20 °C was analyzed by the bending tests with deformation degree of 5.5 % in the spirit with addition of liquid nitrogen in various proportions in order to achieve required temperatures. Results of this test are shown in Fig.4.

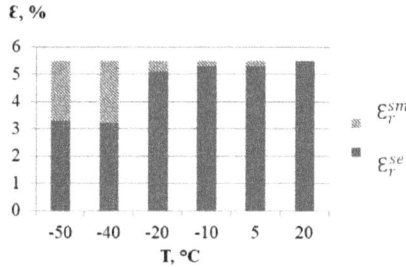

Figure 3. Correlation of superplastic and shape recoverable strain in dependence of deformation temperature during bending tests for rods with a diameter of 3.5 and 4.5 mm.

It can be seen from the diagram that from –30 to 20 °C alloy performs superelastic behavior and from –50 to –30 °C some contribution of shape memory effect appears. The value of shape recovery rate was 100 % from –50 to 20 °C regardless of ratio between shape memory and superelastic effects.

Summary
The rotary forging at 450 °C with post-deformation annealing at deformation temperature allows manufacturing 3 m-long Ti-Ni-Co rods of 3.5 and 4.5 mm in diameter with improved mechanical and functional properties. The obtained rods exhibit superelastic behavior in a wide temperature range –30 to 20 °C with sufficiently high value of a completely recoverable strain of 5 %.

Acknowledgment
The present work was carried out with financial support from the Ministry of Science and Higher Education of the Russian Federation: State Task No. 11.1495.2017/4.6 and State Task No. 007-00129-18-00.

References

[1] Q. Sun, R. Matsui, K. Takeda, E. Pieczyska, Advances in Shape Memory Materials: In Commemoration of the Retirement of Professor Hisaaki Tobushi. Springer, 2017. V.73. https://doi.org/10.1007/978-3-319-53306-3

[2] J. Rui-rui, L. Fu-shun, The influence of Co addition on phase transformation behavior and mechanical properties of TiNi alloys, Ch. J. Aeron. 20 (2007) 153-156. https://doi.org/10.1016/S1000-9361(07)60024-7

[3] E. Sharifi, A. Kermanpur, F. Karimzadeh, The effect of thermomechanical processing on the microstructure and mechanical properties of the nanocrystalline TiNiCo shape memory alloy, Mater. Sci. Eng. A. 598 (2014) 183-189. https://doi.org/10.1016/j.msea.2014.01.028

[4] Y. Kishi, Z. Yajima, K.I. Shimizu, Relation between tensile deformation behavior and microstructure in a Ti-Ni-Co shape memory alloy, Mater.Trans. 43 (2002) 834-839. https://doi.org/10.2320/matertrans.43.834

[5] K. Otsuka, X. Ren, Physical metallurgy of Ti–Ni-based shape memory alloys, Prog. Mater. Sci. 50 (2005) 511-678. https://doi.org/10.1016/j.pmatsci.2004.10.001

[6] S. Prokoshkin, I. Khmelevskaya, V. Andreev, R. Karelin, V. Komarov, A. Kazakbiev, Manufacturing of Long-Length Rods of Ultrafine-Grained Ti-Ni Shape Memory Alloys, In Mater. Sci. F. 918 (2018) 71-76. https://doi.org/10.4028/www.scientific.net/MSF.918.71

[7] A. Lotkov, V. Grishkov, O. Kashin, A. Baturin, D. Zhapova, V. Timkin, Mechanisms of Microstructure Evolution in TiNi-Based Alloys under Warm Deformation and its Effect on Martensite Transformations, Mater. Sci. Found. 81-82 (2015) 245–259. https://doi.org/10.4028/www.scientific.net/MSFo.81-82.245

[8] V. Andreev, V. Yusupov, M. Perkas, V. Prosvirnin, A. Shelest, S. Prokoshkin, I. Khmelevskaya, A. Korotitskii, S. Bondareva, R. Karelin, Mechanical and functional properties of commercial alloy TN-1 semiproducts fabricated by warm rotary forging and ECAP, Russ. Met. (Metally). 10 (2017) 890-894. https://doi.org/10.1134/S0036029517100020

Shape Memory Alloys – SMA 2018   Materials Research Forum LLC
Materials Research Proceedings 9 (2018) 24-27   doi: http://dx.doi.org/10.21741/9781644900017-5

The Effect of the Size Factor on the Functional Properties of Shape Memory Alloy Ring-Shaped Force Elements

Aleksandr Yu. Kiselev[1,a], Nikolay N. Belousov [2],

Elisey A. Khlopkov[3,b], Yuriy N. Vyunenko[4,c*]

[1]JSC ASE EC, 82 Savushkina str., Saint Petersburg 197183, Russia

[2]Donetsk Institute for Physics and Engineering named after A. A. Galkin, 72 R. Luxembourg Str., Donetsk 83114

[3]Peter the Great Saint Petersburg Polytechnical University, 29 Polytechnicheskaya Str., Saint Petersburg 195251, Russia

[4] OOO "OPTIMIKST LTD", 9 Peredovikov Str., Saint Petersburg195426, Russia

[a]aleyukiselev@gmail.com, [b]hlopkovelisey@mail.ru, [c]6840817@mail.ru

*corresponding author

Keywords: Shape Memory, Two-Way Shape Memory, Ring-Shaped Force Elements, Deformation-Force Characteristics, Titanium-Nickel

Abstract. The paper presents the possibility to determine the deformation-force properties of ring-shaped force elements due to the variation of their geometric parameters. It was shown the possibility of using shape memory effect and the two-way shape memory effect in the technological equipment.

Introduction

Use of ring-shaped bundle force elements (RBFE) with shape memory effect (SME) in the design of force generation press type devices "ShER" [1] has demonstrated advantages of application of such devices in technological processes. The research of RBFE has proved the possibility of increasing the efficiency of their characteristics by selecting the SME alloy composition and optimizing their thermomechanical treatment [2]. Ring-shaped band force elements [3] can also serve the same functions in the technological equipment. Thus, to optimize the functional characteristics of the devices one must not only improve the physical and mechanical properties of the alloy, but also the geometric parameters of the force elements. This research presents the results of experiments aimed at determining the role of the ring-shaped force elements diameter on their deformation-force capacity.

Experiment

Rings with a diameter 38.0 mm, 61.5 mm and 79.5 mm were prepared by welding from Ti-50.4 at.% Ni alloy wire with 2 mm diameter. The deformation and force abilities of the obtained "metallic muscles" were determined by a spring dynamometer (Fig. 1)

The rings were heated to 403 K, at which the material of the force elements was in the austenitic state. Then the "metallic muscles" were connected to a helical spring serving as a loading device and a dynamometer, which was the same for all the tests. Then, the assembly of the dynamometer and a ring was transferred from the hot chamber to a refrigerator at a temperature 263 K, which was about 10 K less than M_f – the lower boundary of the direct martensitic transformation temperature interval. During 0.5 hour of holding at this temperature the shape variation of the samples due to the direct transformation was registered. After this, they were again placed into a thermostat with the temperature 403 K. The recovery of the original shape of the ring was monitored during 30 min. Then, the level of the force applied to the

Published under license by Materials Research Forum LLC.

Shape Memory Alloys – SMA 2018   Materials Research Forum LLC
Materials Research Proceedings **9** (2018) 24-27   doi: http://dx.doi.org/10.21741/9781644900017-5

"metallic muscles" was increased and the thermocycle described above was repeated. Four thermal cycles with the initial level of the force 8 N, 14 N, 23 N and 32 N were carried out. The variation of the characteristic ring size $\Delta d = d - d_A$ was fixed, d_A being the ring size in the austenite state, d – its current value. The deformation effects in three samples of different diameters due to the two-way shape memory effect (TWSME) were registered during cooling in the free state in the sixth thermal cycle.

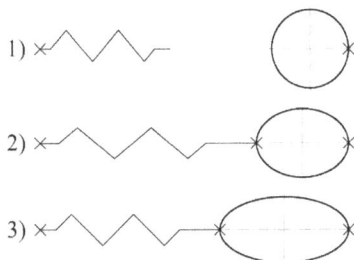

Figure 1. Scheme of the experiment: 1 – ring before loading, 2 – force element after loading in the austenitic state, 3 – sample after the direct martensitic transformation.

To determine the dependence between the deformation-force characteristics of the TWSME and the level of the force acting on the ring-shaped force elements, a series of experiments with rings with a diameter 62 mm were carried out. "Metal muscles" were loaded in the austenitic state. The initial force acting on the first sample was 8 N, on the second – 23 N and on the third – 41 N. Each of the three rings was subjected to more than 25 thermal cycles. The temperature of the hot thermostat was 393–403 K. The temperature in the refrigerator was maintained at 270 K. Stress generation and the deformation effect of TWSME were determined during heating and after cooling the samples in the load-free state.

Fig. 2 shows how the deformation of the ring with the diameter 61.5 mm increases (a) and its force interaction with the spring decreases (b) on cooling in four thermal cycles with different initial load increasing from cycle to cycle. This process is most intensive in the first 15 min while the magnitude of the contact force decreases by 18–25 %. The deformation effect increases with the initial tensile force. On heating because of the SME the accumulated deformation and the level of force interaction of the metallic muscle with the spring are completely restored (Fig. 3) most intensively during about 4 min, which is significantly less than the duration of the deformation increase process on cooling. Apparently, this is a consequence of the fact that the difference between the temperature of the end of the reverse transformation A_f (~ 333 K) and the temperature of the hot thermostat is about 70 K while the M_f temperature is only 10 K above the temperature of the refrigerator. Similar results were obtained for samples of other diameters.

Fig. 4 shows the deformation-force characteristics of all three tested samples. The presented data demonstrate the possibility of obtaining various force effects at a fixed value of Δd and different shape variations under conditions of the same force interaction by changing the diameter of the RBFE.

TWSME appeared as the result of thermocycling of the ring-shaped force elements under load. The deformation Δd of the sample with diameter 38 mm on cooling was 2.3 mm. The deformation effect reached 4.7 mm in the ring with $d_A = 61.5$ mm. Cooling of the force element with the diameter 79.5 mm led to the value of Δd equal to 8.7 mm.

Shape Memory Alloys – SMA 2018   Materials Research Forum LLC
Materials Research Proceedings **9** (2018) 24-27   doi: http://dx.doi.org/10.21741/9781644900017-5

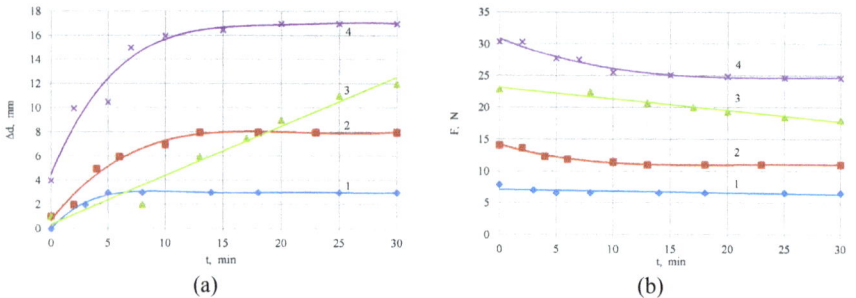

Figure 2. Time dependences of the characteristic ring size Δd (a) and the magnitude
of the force F (b) on cooling with different initial loads: $1 - 8$ N, $2 - 14$ N, $3 - 23$ N, $4 - 32$ N.

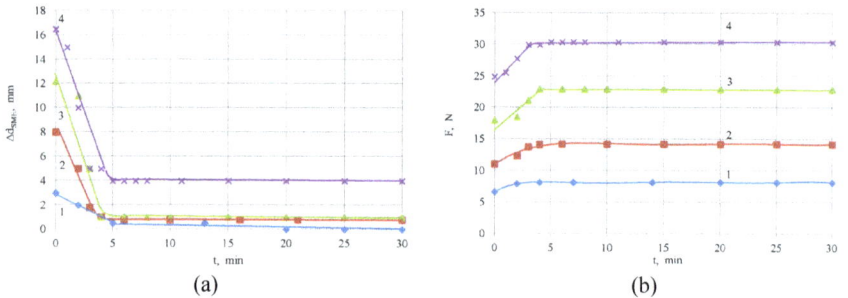

Figure 3. Time dependences of the deformation Δd (a) and force F (b) on heating,
previous cooling having been performed with different initial loads:
$1 - 8$ N, $2 - 14$ N, $3 - 23$ N, $4 - 32$ N.

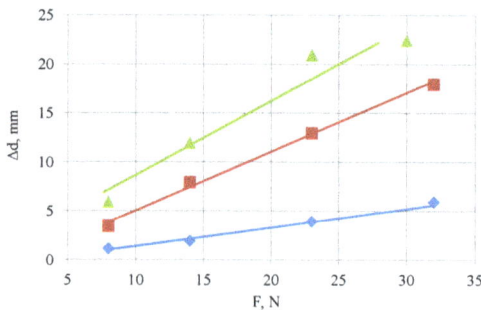

Figure 4. Dependences of the deformation Δd on the initial load F for the rings
with the size d_A 38.0 mm (1), 61.5 mm (2), 79.5 mm (3).

Fig. 5 shows the deformation-force characteristics of the ring upon heating. In this case $\Delta d = d - d_M$, where d_M is the characteristic ring dimension in the martensitic state and d is its current value. The experiments have shown that an increase of the force acting on the rings during thermocycling leads to an increase of the deformation (curve 1, fig. 5) and force characteristics (curve 2, fig. 5) due to TWSME.

Materials Research Forum LLC

doi: http://dx.doi.org/10.21741/9781644900017-5

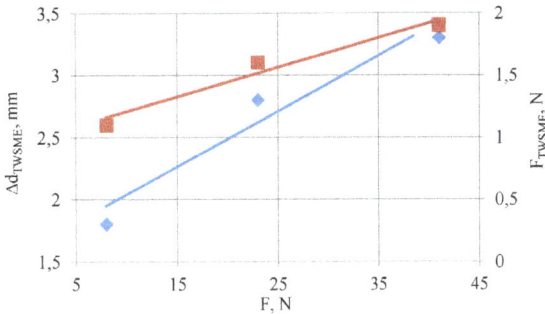

Figure 5. Dependences of the deformation Δd_{TWSME} (1) the generated forces F_{TWSME} (2) on the initial load F.

Conclusion

The obtained results demonstrate the dependence of the deformation-force characteristics of the "metallic muscles" in the form of RBFE on the ring diameter, an increase of which increases its deformability. On the contrary, with a decrease of the "metallic muscles" ring diameter the required level of the force can be reached at a relatively small initial deformation. The kinetics of the deformation variation at operation of an RBFE, and in the conditions of the realization of TWSME are similar. Level of the deformation makes possible to assume that the operation of "SheR" press type devices is suitable for the operation of the "metallic muscles" in both of the two deformation modes: (1) force relaxation on cooling – force generation on heating and (2) deformation in the load-free state due to TWSME on cooling – force generation on heating. The geometric parameters of the metallic muscles can be used as controlling factors in the work of the mechanisms implementing the SME in the technological process. In the development of new devices using RBFE muscles, it is expedient to envisage the option of mounting RBFE with different diameters making the device more universal.

References

[1] Y. Vyunenko, Application of SME in the production of sandwich-type material, in Perspective materials and technologies: collection of articles of International scientific symposium, VSTU Inc., Vitebsk, 2011, pp.182-184.

[2] E. Khlopkov, G. Volkov, Y. Vyunenko, Specific features of the behavior of TiNi force elements in thermocycling, Mater. Today: Proc. 4 (2017) 4879-4883. https://doi.org/10.1016/j.matpr.2017.04.088

[3] Y. Vyunenko, A. Turzakov, E. Khlopkov, G. Volkov, Deformation-force characteristics of ring-shaped bundle force elements from TiNi alloy, in V. Rubanik (Eds.), Perspective materials and technologies: materials of international symposium in 2 parts, Part 1, VSTU Inc., Vitebsk, 2017, pp.36-38.

Shape Memory Alloys – SMA 2018   Materials Research Forum LLC
Materials Research Proceedings 9 (2018) 28-31   doi: http://dx.doi.org/10.21741/9781644900017-6

Influence of Cooling Rate on the Deformation Processes Associated with Direct Martensitic Transformation in TiNi Alloy

Elisey A. Khlopkov[1,a*], Yuriy N. Vyunenko[2,b]

[1]Peter the Great Saint Petersburg Polytechnical University, 29 Polytechnicheskaya Str., Saint Petersburg 195251, Russia

[2]OOO "OPTIMIKST LTD", 9 Peredovikov Str., Saint Petersburg195426, Russia

[a]hlopkovelisey@mail.ru, [b]6840817@mail.ru

*corresponding author

Keywords: Shape Memory, Ring-Shaped Force Elements, Cooling Rate, Deformation Characteristics, Residual Stress Mechanism, Titanium-Nickel

Abstract. The influence of cooling rate on the mechanical characteristics of metals with shape memory is mentioned in connection with the dissipative properties of this material's class. This paper presents the different cooling rate regimes research results of ring-shaped force elements demonstrating the possibility of this parameter influence on their deformation characteristics in the temperature range of direct martensitic transformation.

Introduction

An observation after acting of small-size presses SheR on the shape memory effect (SME) shows the features of the ring-shaped bundle force elements (RBFE, "metal muscles") deformation upon cooling to different temperatures of the heterophase state in the direct martensitic transformation interval [1]. The possibility of using small force effects on "metal muscles" was shown to achieve a significant increase in their deformation-strength properties during the heating as a result of SME. The influence of cooling rate was not fixed on the deformation characteristics. Further studies shows that if final temperatures in the cooling process are below M_f (the temperature of the direct martensitic transformation finish), then the deformation processes depend from the cooling rate.

Experiment

The research was taking place under ring-shaped single-turn force elements. The samples were prepared from TiNi50.4 at.% alloy wire with 2 mm diameter. The possibility of preparing the force elements were considered for operation due to their deformation under the condition of the direct transformation plasticity evolution in the material.

A force element having the shape of an elongated oval (Fig.1, 1) was loaded along its large diameter axis during the experiment. In the initial state, this value d_1 was equal to 73 mm. A force of ~ 6.4 N was applied at a thermostat where the temperature was 400 K. The simplest dynamometer spring was used (Fig.1, 2) for this purpose. After that, cooling was done. During cooling with a thermostat, the reference dimension d_1 increased by 7.5 mm. At the same time, the force of interaction with the spring of the dynamometer decreased to 2.2 N. This is demonstrated qualitatively in Fig.1, 3.

During the first 100min of cooling, the shape of the force element did not change. The increase in d_1 begins when the temperature reaches ~ 320 K. The temperature of the thermostat decreases by this minute into this level. Then, after 200 min, the deformation processes end at ~ 300 K (Fig.2, curve 1).

Published under license by Materials Research Forum LLC.

Shape Memory Alloys – SMA 2018   Materials Research Forum LLC
Materials Research Proceedings 9 (2018) 28-31   doi: http://dx.doi.org/10.21741/9781644900017-6

When the dynamometer with the power element is extracted from the hot thermostat and cooled in the air at room temperature, the oval change becomes noticeable after 8 min. The deformation process continues for ~ 60 min. There is a change in d_1 to 6.5 mm, which is 13 % less than the corresponding value in the first experiment. The force of interaction with the elastic counterbody decreased to ~ 2.75 N (Fig. 2, curve 2).

If the dynamometer with an oval is transferred from the thermostat to the refrigerators with temperatures ~ 280 K and ~ 270 K, then the oval shaping begins very quickly (during the first 4 minutes in the first case (Fig.2, curve 3) and after 3 minutes in the second case (Fig.2, curve 4)). In isothermal holding temperature of 280 K d_1 increases only by 5mm and by 3.5 mm at 270 K. Accordingly, the force effect is reduced to 3.4 N, and to 4.4 N.

The heating of the deformed force elements led to a complete shape memory in all four experiments.

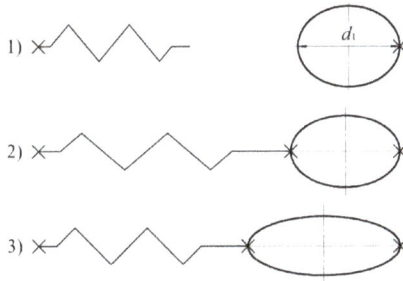

Figure 1. Scheme of the experiment: 1 – ring before loading, 2 – force element after loading in the austenitic state, 3 – sample after the direct transformation plasticity.

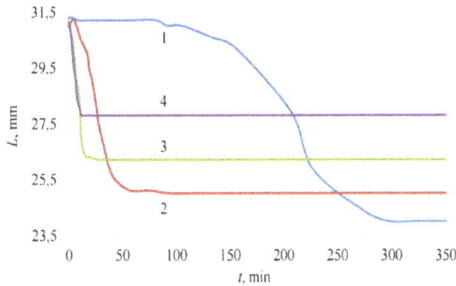

Figure 2. The influence of cooling rate on TiNi oval shaping: 1 – cooling with a thermostat; isothermal: 2 – at 300 K, 3 – 280 K, 4 – 270 K.

The numerical experiments were done to analyze the possible causes of the cooling rate influence. Calculations can be made using the model described in [2], but the authors take the heat capacity of the material to be a constant. In the framework of the residual stress mechanism [3], the temperature distribution in the cross section of the cylinder was calculated upon cooling it under isothermal conditions when the surface temperature was equal to the lower limit of the M_f conversion and at a temperature 40 K below M_f. For the calculations, the heat equation in cylindrical coordinates (1) is chosen:

$$\rho \cdot c(U) \cdot \frac{\partial U}{\partial t} = k \cdot \left(\frac{\partial^2 U}{\partial r^2} + \frac{1}{r} \cdot \frac{\partial U}{\partial r} \right),$$

(1)

where ρ – material density; k – thermal conduction coefficient; $c(U)$ – heat capacity, U – temperature, r – radial coordinate, t – time. $c(U) = c_0$ is a constant in the single-phase state. Taking into account the calorimetric research results assumed by the numerical experiment that the specific heat values in the martensitic and austenite state are close in magnitude. Under the transformation the temperature dependence of heat capacity were approximated by the quadratic function (2):

$$c(U) = c_1 \cdot \frac{(U - M_s) \cdot (M_f - U)}{(M_f - M_s)^2} + c_0,$$

(2)

where M_s – temperature of the direct martensitic transformation start, M_f – temperature of the direct martensitic transformation finish. The coefficient c_1 is determined from equation (3):

$$Q_{tr} = \int_{M_s}^{M_f} c_1 \cdot \frac{(U - M_s) \cdot (M_f - U)}{(M_f - M_s)^2} dU,$$

(3)

where Q_{tr} – latent heat of transformation.

Fig. 3 shows curves which are demonstrating the evolution of the temperature field at the surface temperature $U_{surf} = M_f - 40$ (Fig.3, a) and $U_{surf} = M_f$ (Fig.3, b). The model constants were chosen to be close to the corresponding values of the equiatomic TiNi alloy. The results point out that in the case of the cylinder surface subcooling, the heterophase state does not cover the entire volume of the material. Because of the strong lowering of the surface temperature (~ 130 K), the martensitic transformation begins in the near-surface layer and goes in a narrow band. The internal volume of the material to 0.9r (r – radius of the cylinder) is still in the austenitic state, but the alloy surface is already in the martensitic state (Fig.3, a, curve 111,7). Gradually the heterophase state covers the entire inner region of the cylinder. Simultaneously, the martensitic ring becomes wider. Thus, a significant part of the material in the volume structure at any time is outside the conditions of the direct transformation plasticity. In this case, the phase transformation throughout the volume of the material is completed approximately 8 times faster than in the second case (Fig.4). Calculations demonstrated that the surface subcooling at 40 K increases the cooling rate of the material in the cylinder inner regions 1.5–2 times, and in the near-surface layer the cooling rate is 5–6 times higher. This restrict the evolution of deformation processes corresponding to the laws of high-speed creep, which characterize the heterophase state, i.e. deformation phenomena of the direct transformation plasticity.

At the temperature of the surface M_f, the temperature field varies smoothly over the volume of the cylinder (Fig.3, b). Therefore, the heterophase state covers the entire volume of the cylinder (Fig.4, b). The material stays relatively long in this state. This makes the construction more plastic, which leads to an increase in the shape change at equal forces applied to the "metal muscles" and their cooling through the interval of a direct martensitic transformation.

Shape Memory Alloys – SMA 2018   Materials Research Forum LLC
Materials Research Proceedings **9** (2018) 28-31   doi: http://dx.doi.org/10.21741/9781644900017-6

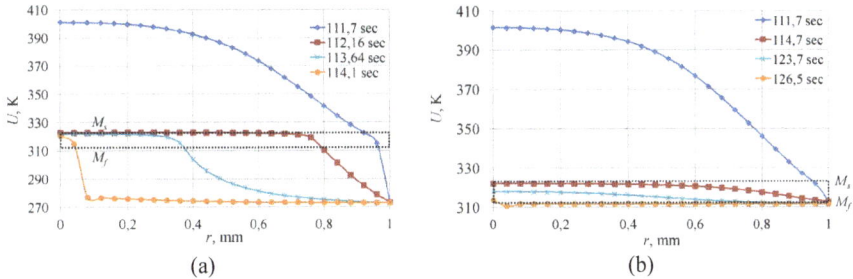

Figure 3. Evolution of the temperature field in the cylinder r = 1 mm with a sudden change in the surface temperature from 400 K to M_f – 40 K (a), to M_f (b).

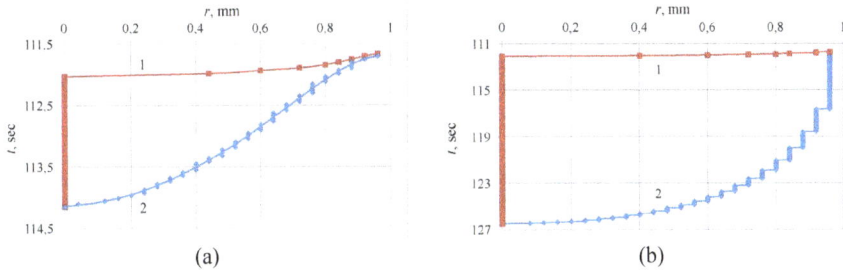

Figure 4. Evolution of the zone of direct martensitic transformation in the cylinder r = 1mm over time with a sudden change in the surface temperature from 400K to M_f – 40K (a), to M_f (b): 1 – boundary of the direct transformation beginning, 2 – the end of the direct conversion.

Conclusion

Results of the force elements research and the numerical experiments have shown the importance of cooling rate in the organization of deformation processes in the material during direct martensitic transformation. This factor can be used in the technology of applying mechanisms acting on the shape memory effect as a regulator of their deformation-force characteristics.

References

[1] E. Khlopkov, G. Volkov, Y. Vyunenko, Specific features of the behavior of TiNi force elements in thermocycling, Mater. Today: Proc. 4 (2017) 4879-4883. https://doi.org/10.1016/j.matpr.2017.04.088

[2] A.E. Volkov, A.S. Kukhareva, Calculation of the stress-strain state of a TiNi cylinder subjected to cooling under axial force and unloading, Bull. Russ. Acad. Sci. Phys. 72 (2008) 1267-1270. https://doi.org/10.3103/S106287380809027X

[3] Yu.N. Vyunenko, The mechanism of shape memory effect caused by the residual stress field evolution, Materialovedenie. 12 (2003) 2-6.

Shape Memory Alloys – SMA 2018
Materials Research Proceedings 9 (2018) 32-37

Materials Research Forum LLC
doi: http://dx.doi.org/10.21741/9781644900017-7

Thermomechanical and Magnetic Properties of Fe-Ni-Co-Al-Ta-B Superelastic Alloy

Victor V. Koledov[1,a], Elvina T. Dilmieva[1,b], Vladimir S. Kalashnikov[1,c],
Alexander P. Kamantsev[1,d*], Alexey V. Mashirov[1,e], Svetlana V. von Gratowski[1],
Vladimir G. Shavrov[1], Alexey V. Koshelev[2], Vedamanickam Sampath[3],
Irek I. Musabirov[4], Rostislav M. Grechishkin[5]

[1]Kotelnikov Institute of Radio-engineering and Electronics of RAS, 11-7 Mokhovaya Str.,
Moscow 125009, Russia

[2]Lomonosov Moscow State University, 1 Leninskie Gory, Moscow 119991, Russia

[3]Indian Institute of Technology Madras, Chennai 600036, India

[4]Institute for Metals Superplasticity Problems of RAS, 39 Stepana Halturina Str.,
Ufa 450001, Russia

[5]Tver State University, 33 Zhelyabova Str., Tver 170100, Russia

[a]victor_koledov@mail.ru, [b]kelvit@mail.ru, [c]vladimir.kalashnikovS@gmail.com,
[d]kaman4@gmail.com, [e]a.v.mashirov@mail.ru

*corresponding author

Keywords: Phase Transition, Shape Memory Effect, Superelasticity, Strain Glass, Spin Glass, Fe-Ni-Co-Al-Ta-B, Superstructures

Abstract. The ingot of $Fe_{40.71}Ni_{27.33}Co_{17.13}Al_{12.05}Ta_{2.73}B_{0.05}$ alloy was produced by arc melting technique followed by heat treatment. The alloy ingot was cut by electro-discharge machining and was further subjected to rolling. The microstructure of surface, thermomechanical and magnetic properties were studied. The alloy exhibits superelasticity at temperature lower than 330 K. The hysteretic behavior of magnetization was observed. These properties can be explained by combination of states of the spin- and strain-glasses.

Introduction

Ferromagnetic iron-containing Heusler alloys exhibiting shape memory effect (SME), such as Ni-Mn-Ga-Fe demonstrate reversible deformations of up to several % due to thermoelastic martensitic transition induced by magnetic field [1]. The iron-based, high-strength Fe-Ni-Co-(Al-Ta-B) alloy with superplasticity and reversible deformations of more than 13 % with a tensile strength above 1 GPa is discussed in [2]. The tensile strength of this alloy is almost twice that for the highest stress for superelastic deformation in Ni-Ti alloys. In addition, this iron-containing alloy also shows high damping and reversibility of magnetization during loading and unloading processes. More recently, other iron-containing alloys exhibiting superelasticity have attracted attention [3-7]. The investigations of a nanostructured FeMnSi shape memory alloy produced via severe plastic deformation is done in [8]. The family of Fe-Ni-Co-(Al-Ta-B) alloys demonstrates unique physical properties: a wide temperature hysteresis of superelasticity, a change in magnetization over a larger range, and electrical resistance due to load, accompanied by high pseudoplasticity. In view of these properties, these iron-containing superelastic alloys are expected to be used for a wide range of practical applications, such as damping and functional materials. Significant interest in these alloys is caused by the quest for inexpensive structural materials for the fabrication of structures that are resistant to earthquakes, such as nuclear power plants, high-rise buildings, bridges and industrial facilities.

Published under license by Materials Research Forum LLC.

This paper presents the results of the experimental study of new $Fe_{40.71}Ni_{27.33}Co_{17.13}Al_{12.05}Ta_{2.73}B_{0.05}$ (FNCATB) alloy: the microstructure of surface, the occurrence of structural phase transitions (PT) due to temperature, mechanical stress as well as magnetic field in wide temperature range 4–400 K.

Samples and Microstructure

The FNCATB alloy ingot was made by arc melting in an argon atmosphere with subsequent heat treatment. The ingot was annealed for 24 hours at temperature of 1493 K, followed by a thermal aging of 72 hours at 873 K. Then the plates were cut by the electroerosive method in the form of plates, which were subsequently thinned by cold rolling.

The homogeneity of the FNCATB alloy was investigated by the EDX method – the chemical composition of the sample was studied at 8 points. Table 1 presents the maximum and minimum values of the weight percent of the chemical elements fixed in the alloy in the volume under study. The obtained alloy has a satisfactory homogeneity and the obtained chemical composition is close to the one set within the error of measurement. The EDX method has an error of about 0.3–0.5 % by weight, therefore it does not allow to fix the presence of boron in the sample.

Table 1. The composition of the sample of the FNCATB alloy [at.%].

	Al	Fe	Co	Ni	Ta	B
Max	6.98	40.19	17.53	28.46	6.61	-
Min	6.26	38.80	17.10	27.62	7.47	-

Studies of the microstructure of the sample using a scanning electron microscope revealed the multiphase nature of the FNCATB alloy at room temperature. Also, it can be seen from Fig. 1(a), that the sample has a dendritic structure, possibly associated with the heat sink process in the manufacture of the sample.

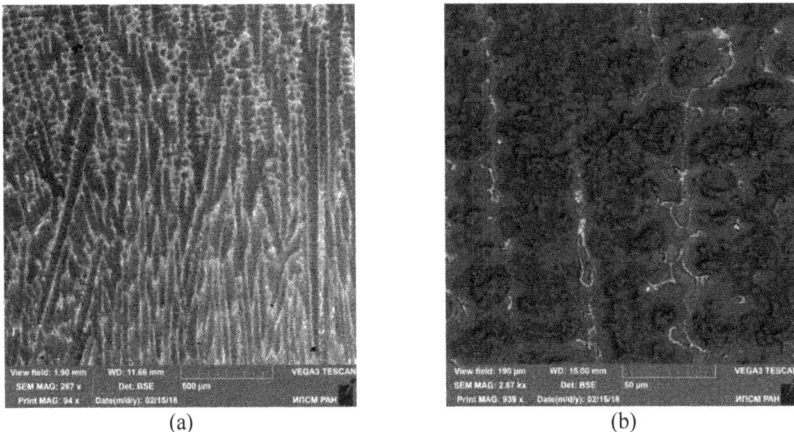

(a)   (b)

Figure 1. Photographs of the surface structure of the FNCATB sample taken with a scanning electron microscope: (a) 500 μm and (b) 50 μm scales.

Metallographic studies were carried out by polarization optical microscope. The sample was polished above room temperature at 330 K. Fig. 2 (a) demonstrates the microstructure of the FNCATB alloy at room temperature. The martensitic phase appeared on the surface of the metallographic section after additional thermal cycling (cooling down to K and heating to

Shape Memory Alloys – SMA 2018
Materials Research Proceedings **9** (2018) 32-37

Materials Research Forum LLC
doi: http://dx.doi.org/10.21741/9781644900017-7

300 K), which also proves the presence of the structural PT of the 1-st order in the FNCATB alloy (Fig. 2(b)). The additional electropolishing (Fig. 2(c)-(f)) made it possible to reveal the nature of the precipitates of the phases of the alloy.

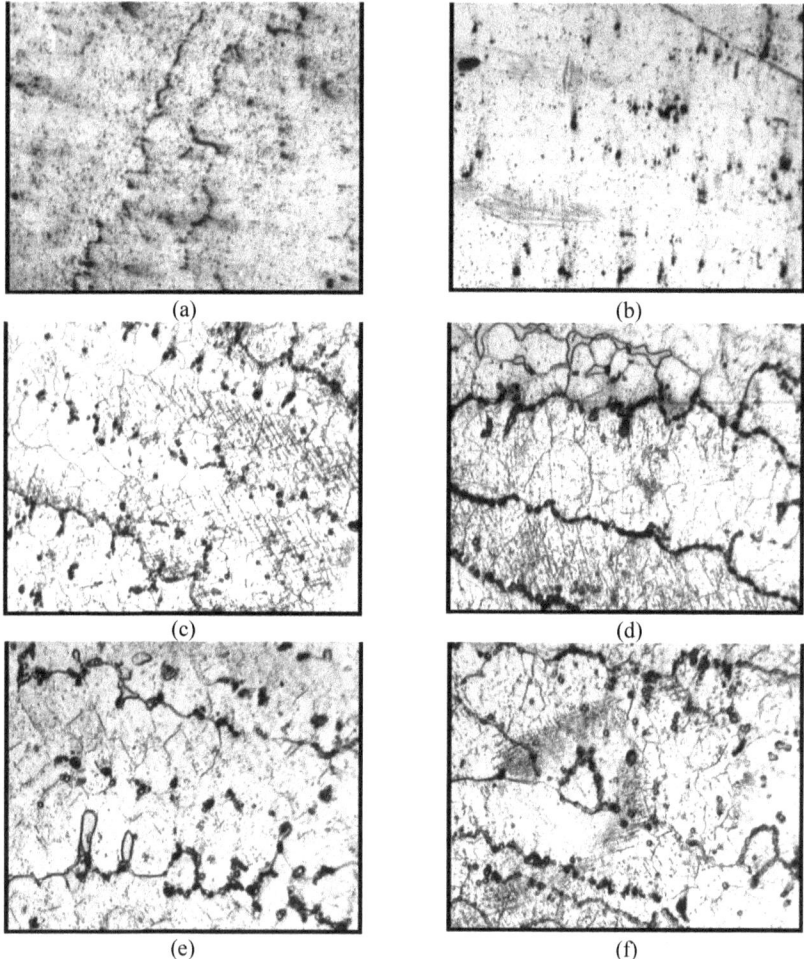

(a)

(b)

(c)

(d)

(e)

(f)

Figure 2. The microstructure of the FNCATB (a) alloy at room temperature, (b) after additional cooling downto 77 K and heating to 300 K, (c)–(f) after electropolishing at room temperature.

Experimental Results

The Fig. 3 shows the temperature dependences of the magnetization of the FNCATB sample obtained by using the ZFC-FC-FH protocol in magnetic field with induction of 3 T and in low field of 50 Oe (in the inset). There is an anomaly of the behavior of the magnetization curve like hysteresis in low field near the room temperature, which presupposes the presence of the 1-st order PT in the alloy. The magnetization dependence typical for ferromagnetic materials was

Shape Memory Alloys – SMA 2018   Materials Research Forum LLC
Materials Research Proceedings **9** (2018) 32-37   doi: http://dx.doi.org/10.21741/9781644900017-7

observed in the high field during the heating, hysteresis behavior was observed during the cooling. These results can be explained by the combination of states of the spin and deformation glass in the FNCATB alloy in wide temperature range of 100–400 K.

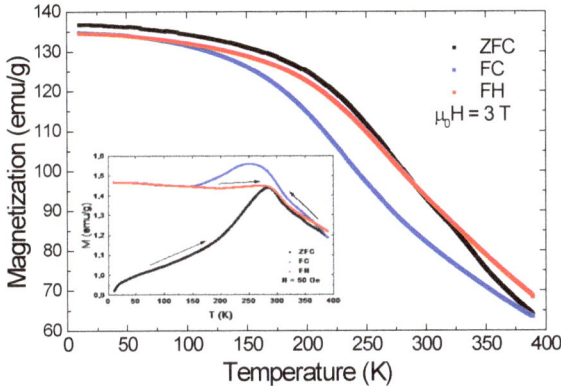

Figure 3. Temperature dependence of magnetization of the FNCATB alloy in magnetic field with induction of 3 T. In the inset: the temperature dependence of magnetization of the FNCATB alloy in magnetic field of 50 Oe.

To study the functional properties of alloys with giant effect of superelasticity an experimental setup was used to determine the thermomechanical properties of alloys in the temperature range 140–570 K, mechanical stresses up to 2000 MPa and deformations up to 20 %. The principle of the installation is based on the method of three-point bending of the sample at variable temperature and constant load. The installation was tested on samples of the known alloy $Ni_{49.8}Ti_{50.2}$, which confirmed the reliability of the results obtained on it [9].

The Fig. 4 shows the results of the study of the temperature dependence of the deformation of the FNCATB sample. The research protocol was as follows. The sample was heated from 140 K to 370 K without load. Then, the stresses of 527 MPa were reached by applying external mechanical force at temperature of 370 K. The sample was cooled to 140 K and then heated to 410 K under load. The deformation versus temperature curve for heating and cooling shows deviation from the linear dependence, which indicates the presence of pseudoplasticity and SME in the sample. The value of the SME is not more than 0.3%. The temperatures of the beginning and the end of the thermoelastic PT are indicated in the Fig. 4.

The Fig. 5 shows the dependence of the strain of the sample upon loading at 368 K. The dependence also reflects the deviation from Hooke's law (from linearity). The curve σ (ε) shows the location of the kink, where a deviation is observed, which corresponds to the stress of the PT under load and is approximately 370 MPa.

The Fig. 6 shows the dependence of the deformation on the load for three loading-unloading cycles at temperature of 298 K. It is noted that as the number of thermocycles increases, the stress at which the PT is observed decreases. On the graphs, when the load is removed, the deformation of the sample after the load is removed is somewhat lower than the deformation of the sample under load. This phenomenon is instrumental in nature and is a feature of the installation on which the test was carried out.

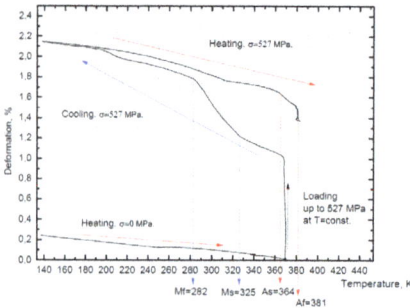

Figure 4. Temperature dependence
of the deformation of the FNCATB alloy
under different loads.

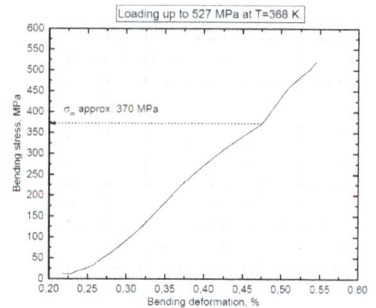

Figure 5. Dependence of the deformation
of the FNCATB alloy under loading
at 368 K.

The Fig. 7 shows the results of thermocycling the sample under load and without load. The structural PT was observed during the first cycles of heating loading in the FNCATB sample. The load was removed on the third cycle, and the PT was observed before the 7th cycle. Then the PT was absent – the effect of the sample training disappeared or the relaxation of the stresses induced during the first reward cycles occurred.

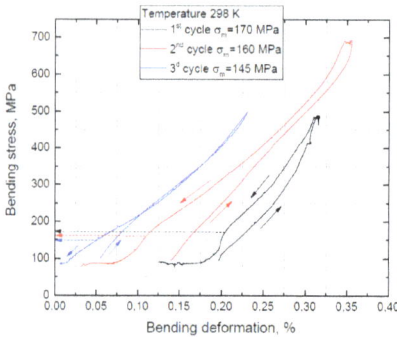

Figure 6. Dependence of strain on load
for three loading-unloading cycles
of the FNCATB alloy.

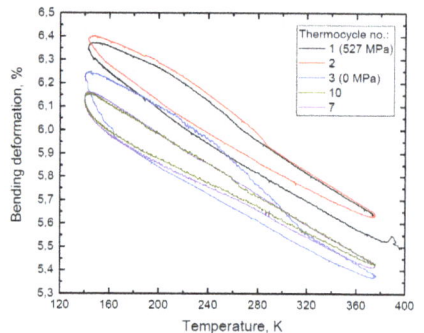

Figure 7. Temperature dependence
of the deformation of the FNCATB alloy
under load and without load
up to the 7th thermal cycle.

Conclusions

In this paper we present the results of the study of the $Fe_{40.71}Ni_{27.33}Co_{17.13}Al_{12.05}Ta_{2.73}B_{0.05}$ (FNCATB) alloy, an experimental study of their microstructure, the appearance of a structural PT induced by temperature and mechanical stress, and also by a magnetic field.

1) The ingot of FNCATB alloy was made by the arc melting method followed by homogenizing annealing and ageing, then plates were made by electric cutting and rolling. The initial composition was confirmed by EDX analysis.

Shape Memory Alloys – SMA 2018   Materials Research Forum LLC
Materials Research Proceedings **9** (2018) 32-37   doi: http://dx.doi.org/10.21741/9781644900017-7

2) The microstructure of the alloy surface was studied by optical metallography and scanning electron microscopy. The multiphase of the FNCATB alloy was revealed at room temperature, and a dendritic surface structure was also detected.

3) The typical behavior of magnetization versus temperature dependence at heating is observed, and hysteresis behavior is manifested at cooling in high magnetic field. These results can be explained by the combination of states of the spin- and strain-glass in the FNCATB alloy in wide temperature range 100–400 K.

4) The thermoelastic PT of the 1-st order was detected by the dilatometric method in the temperature range 282–381 K, while the SMA value does not exceed 0.3 % at load of 527 MPa. The effect of superelasticity at temperatures below 330 K in the FNCATB alloy is shown.

Studies of iron-containing alloys with effects of superelasticity should be continued, because a combination of strong superelasticity and high reliability with a relatively low price and manufacturability is of great interest for creating superstructures resistant to extreme loads.

Acknowledgments

The work is supported by RFBR, Grants No. 16-57-45066, 17-07-01524.

References

[1] A.A. Cherechukin et al., Shape memory effect due to magnetic field-induced thermoelastic martensitic transformation in polycrystalline Ni-Mn-Fe-Ga alloy. Phys. Lett. A. 291 (2001) 175. https://doi.org/10.1016/S0375-9601(01)00688-0

[2] Y. Tanaka et al., Ferrous polycrystalline shape-memory alloy showing huge superelasticity, Science. 327 (2010) 1488. https://doi.org/10.1126/science.1183169

[3] S. Bhowmick, S.K. Mishra, FNCATB Superelastic damper for seismic vibration mitigation, J. Intel. Mat. Syst. Str. 27 (2016) 2062. https://doi.org/10.1177/1045389X15620039

[4] T. Omori, K. Ando, M. Okano, X. Xu, Y. Tanaka, I. Ohnuma, K. Ishida, Superelastic effect in polycrystalline ferrous alloys, Science. 333 (2011) 68-71. https://doi.org/10.1126/science.1202232

[5] T. Omori, S. Abe, Y. Tanaka, D.Y. Lee, K. Ishida, R. Kainuma, Thermoelastic martensitic transformation and superelasticity in Fe-Ni-Co-Al-Nb-B polycrystalline alloy, Scripta Mater. 69 (2013) 812-815. https://doi.org/10.1016/j.scriptamat.2013.09.006

[6] D. Lee, T. Omori, R. Kainuma, Ductility enhancement and superelasticity in Fe-Ni-Co-Al-Ti-B polycrystalline alloy, J. Alloy. Compd. 617 (2014) 120-123. https://doi.org/10.1016/j.jallcom.2014.07.136

[7] Y. Tanaka, R. Kainuma, T. Omori, K. Ishida, Alloy Design for Fe-Ni-Co-Al-based Superelastic Alloys, Mater. Today: Proc. 2 (2015) S485-S492. https://doi.org/10.1016/j.matpr.2015.07.333

[8] G. Gurau, C. Gurau, V. Sampath, L.G. Bujoreanu, Investigations of a nanostructured FeMnSi shape memory alloy produced via severe plastic deformation, Int. J. Miner. Metall. Mater. 23 (2016) 1315-1322. https://doi.org/10.1007/s12613-016-1353-6

[9] V.S. Kalashnikov, V.V. Koledov, D.S. Kuchin, A.V. Petrov, V.G. Shavrov. A three-point bending test machine for studying the thermomechanical properties of shape memory alloys, Instrum. Exp. Tech. 61 (2018) 306-312. https://doi.org/10.1134/S0020441218020148

Shape Memory Alloys – SMA 2018   Materials Research Forum LLC
Materials Research Proceedings 9 (2018) 38-42   doi: http://dx.doi.org/10.21741/9781644900017-8

Structure Formation, Mechanical and Functional Properties of Ti-Ni SMA, Deformed by Compression in a Wide Temperature Range

Irina Yu. Khmelevskaya[1,a], Viktor S. Komarov[1,2,b*,] Ivan A. Postnikov[1,c],
Roman D. Karelin[1,d], Grzegorz Korpala[2,e], Rudolf Kawalla[2,f],
Sergey D. Prokoshkin[1,g]

[1]National University of Science and Technology "MISiS", 4 Leninskiy Prospect, Moscow 119049, Russia

[2]Technische Universität Bergakademie Freiberg, 4 Bernhard-von-Cotta-Str., Freiberg D-09599, Germany

[a]khmel@tmo.misis.ru, [b]komarov@misis.ru, [c]postnikov.ivan@rambler.ru, [d]rdkarelin@gmail.com, [e]grzegorz.korpala@imf.tu-freiberg.de, [f]rudolf.kawalla@imf.tu-freiberg.de, [g]prokoshkin@tmo.misis.ru

*corresponding author

Keywords: Shape Memory Alloys, Ti-Ni, Structure Formation, Aging, Dynamic Softening

Abstract. The structure, mechanical and functional properties of the aging Ti-50.8 at.% Ni shape memory alloy after deformation by compression in a wide temperature range of 100–900 °C and post-deformation annealing at 430 °C was studied. The temperature regions of dynamic softening processes (recovery, polygonization and recrystallization) were determined as a result of comparative study (TEM, X-ray, changes in properties). It was found that dispersion hardening due to precipitation of Ti_3Ni_4 phase particles is accompanied by an increase in functional properties.

Introduction

Expanding the scope of shape memory alloys (SMA) application and increasing the complexity of medical devices using shape memory effect (SME) lead to stricter of requirements for the functional properties (FP) of Ti-Ni-based SMA. At the present time the most promising way to control the FP is severe plastic deformation (SPD), creating ultrafine-grained structure up to nanocrystalline structure [1]. The refinement of the structure is usually achieved by increasing of accumulated strain and lowering of the deformation temperature [2]. For obtaining of ultrafine-grained (submicrocrystalline) structure in Ti-Ni alloys different schemes of SPD are used: equal channel angular pressing [3], rotary forging [4], MaxStrain-deformation [5,6] and high pressure torsion [7]. For further testing of different SPD modes at lower temperatures (below 400 °C), it is necessary to know the temperature ranges of dynamic processes of hardening and softening. In addition, in nickel-rich Ti-Ni alloys for medical purposes aging process take place during deformation and/or post-deformation annealing (PDA). Precipitation of Ti_3Ni_4 phase particles affects the structure and properties of Ti-Ni SMA [8].

Thus, the optimization of technological processes of production Ti-Ni SMA with the high level of FP requires more detailed knowledge of the structure formation in a wide range of deformation temperatures. Therefore, the aim of this work was searching for optimal regimes of combined thermomechanical treatment (deformation + post-deformation annealing) for Ti-Ni SMA manufacturing.

Published under license by Materials Research Forum LLC.

Shape Memory Alloys – SMA 2018
Materials Research Proceedings **9** (2018) 38-42

Materials Research Forum LLC
doi: http://dx.doi.org/10.21741/9781644900017-8

Experimental

Plastometric compression tests of the aging Ti-50.8 at.% Ni alloy were performed on the hot deformation simulator "WUMSI" at a deformation rate of 1 s^{-1} in the temperature range from 100 to 900 °C on samples 5 mm in diameter and 10 mm in height witch have after the reference treatment (700 °C, 30 min followed by cooling in water), the following characteristic temperatures of the martensitic transformations B2 $\leftrightarrow$ B19': $M_s = -82$, $M_f = -96$, $A_s = -64$, $A_f = -44$ °C. The structure, substructure and phase composition after deformation by compression at a given temperature with accumulated strain $e = 0.5$, quenching and subsequent annealing at 430 °C, 1 hour, was examined by X-ray structural analysis, light and transmission electron microscopy. PDA at 430 °C is usually used to eliminate excessive strain hardening and increase plasticity. It helps to achieve the required functional properties and is necessary to set the shape of the final product. The selected annealing temperature, 430 °C, according to works [8,9], corresponds to the most intensive aging. The mechanical properties were evaluated by Vickers hardness tests. The functional properties (the value of a completely recoverable strain) were evaluated by a thermomechanical method, including deformation by bending in liquid nitrogen and heating to restore the shape.

Results and discussion

Mechanical properties. Lowering the deformation temperature from 900 to 100 °C is accompanied by an increase in hardness from 280 to 350 HV, which indicates an increase in structure defects. At a deformation temperature of 600 °C and higher, the hardness approaches the level of the reference treatment, which indicates the progress of the recrystallization processes (Fig. 1). In contrast to the equiatomic alloy in an "entangled" alloy, the PDA at 430 °C is accompanied by an increase in hardness after deformation in the range 100–600 °C. This PDA effect is an obvious consequence of the static dispersion hardening and retardation of softening pre-recrystallization processes by Ti_3Ni_4 phase particles released during aging of alloys, which was confirmed by structure studies (Fig. 1).

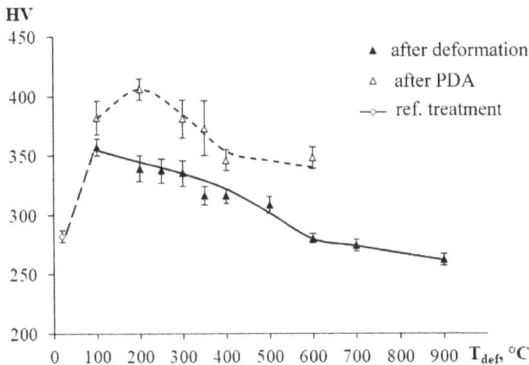

Figure 1. Vickers hardness vs deformation temperature and post-deformation annealing.

Structure formation. Light microscope study of the grain structure of the alloy reveals that the average grain size of B2-austenite, measured after deformation at temperatures from 100 to 600 °C, increases slightly with increasing in deformation temperature from 16 to 22 μm, while the elongated shape of the grain is preserved. Deformation at 700 °C leads to the formation of new fine grains (due to dynamic recrystallization), while the average grain size does not practically change (23 μm after reference treatment). An increase in the deformation temperature to 900 °C leads to complete recrystallization and an increase in the average grain size to 33 μm [10].

Electron microscopic study of the structure was observed immediately after deformation by compression at temperatures of 200, 400, 600 °C and after PDA [10]. The results are given only for deformation at 200 °C and PDA at 430 °C, since they demonstrate both cases: the initial one without the precipitates of the Ti_3Ni_4 phase and aged as a result of the PDA.

Deformation at a temperature of 200 °C leads to the formation of a developed dislocation substructure with a very high dislocation density in the background of deformation bands and / or martensitic-like approximately parallel plates (Fig. 2a). Reflexes of B2-austenite and numerous reflexes of R-phase are identified on the microdiffraction pattern. The characteristic phase reflexes are indicated by arrows. There are no explicit traces of the Ti_3Ni_4 phase. They are identified after the PDA at 430 °C and are accompanied by the appearance in the bright- and dark-field images of contrast in the form of an extremely dispersed (nano-sized) point ripple that veils the dislocation substructure and shadows the boundaries of deformation bands in B2-austenite and R-phase crystals. (Fig. 2 b). In the process of deformation at a temperature of 400 °C, dynamic deformation aging develops. After the PDA, all three phases are present, with particles of the Ti_3Ni_4 phase growing noticeably. The polygonized dislocation substructure after this treatment differs with difficulty, due to superposition of R-phase images and Ti_3Ni_4 particles released during aging. After deformation at a temperature of 600 °C, fairly large grains with a small dislocation density (about 10^9 cm^{-2}) are observed. In the structure there are B2-austenite and R-phase, and no signs of the presence of the Ti_3Ni_4 phase were detected. PDA leads to separation of particles of the Ti_3Ni_4 phase larger than those after deformation at lower temperatures [10].

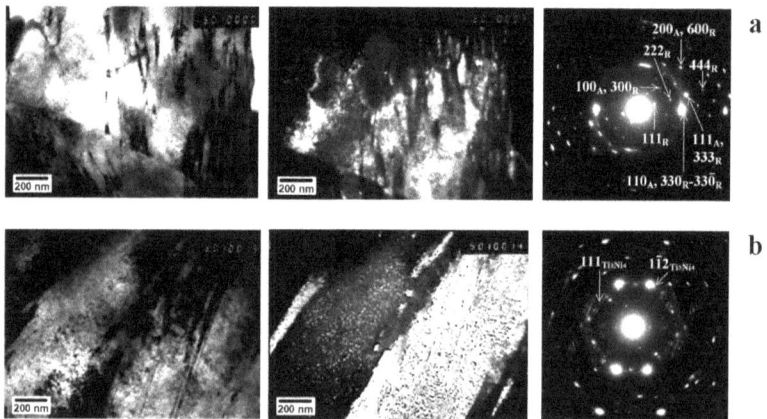

Figure 2. Structure of Ti-50.8 at.% Ni alloy: (a) after deformation at T = 200 °C; (b) after PDA at 430 °C.

X-ray diffraction. On the results of X-ray diffraction analysis, the {110} B2-austenite X-ray line width of was plotted against the deformation temperature (Fig. 3). With deformation temperature increasing, it decreases at first slowly (up to 300 °C), then rapidly (from 300 to 600 °C) and then stabilizes (Fig. 3). Such a change in the width of the line indicates a decrease in the degree of crystal lattice defectness due to the successive flow of recovery, polygonization, and recrystallization processes. The PDA is accompanied by some narrowing of the B2-austenite lines – the defectness of the lattice decreases.

PDA at a temperature above the deformation temperature leads to a decrease in the defectness of the B2-austenite lattice and slightly reduces the width of the X-ray lines (Fig. 3).

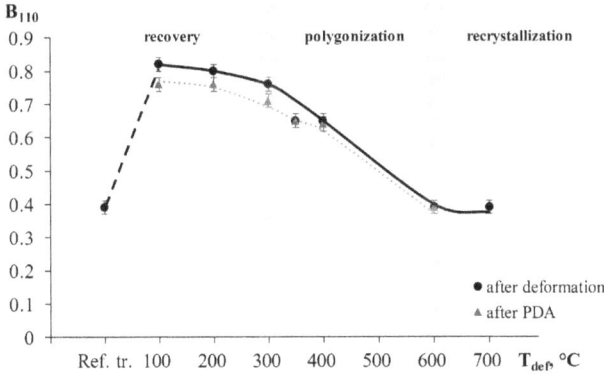

Figure 3 Dependence of the {110} B2-austenite X-ray line width on deformation temperature.

Comparative study of Ti-Ni SMA structure, phase formation and changes in hardness in a wide range of deformation temperatures (100–900 °C) allow to determine the boundaries of the temperature regions of dynamic softening processes under conditions of deformation (Fig. 3). The dynamic recovery region of the alloy is 100–300 °C, which follows from the observation of a very high density of dislocations and a honeycomb type of substructure, as well as the prevalence of dynamic strain hardening over softening. The region of dynamic polygonization of the alloy is 300–600 °C, as evidenced by the formation of the polygonized substructure, the accelerated decrease in the width of the X-ray line, and the attainment of the steady-state stage in the flow curves [10]. The region of dynamic recrystallization lies above 600 °C due to new recrystallized grains and the "return" of the width of the X-ray line to the level of reference treatment processing.

Functional properties. Alloy's ability to shape recovery was studied when deformation was induced in liquid nitrogen and was estimated from the value of total recoverable strain ε_{rt} due to the realization of the memory effect (SME) and superelasticity (SE), i.e. in magnitude: $\varepsilon_{rt} = \varepsilon_{rt}^{SME} + \varepsilon_{r}^{SE}$. After all the deformation modes, the value of the completely recoverable strain was not less than 7 %. The maximum values $\varepsilon_{rt} = 9$ % were obtained after deformation at 400–600 °C. The PDA led to an increase in ε_{rt} in the case of deformation at 400 °C and higher – to a maximum value of 10–11.5%. A change in the ratio of the components of the total recoverable strain ε_{rt} from the deformation modes was established. The ratio of the contributions of the SME and SE to the total recoverable strain after most regimes was in favor of the SME. After deformation at 300 °C, these fractions were close in magnitude ($\varepsilon_{r}^{SE} = 6.4$ % and $\varepsilon_{r}^{SME} = 5.4$ %).

Summary

1. The temperature regions of the development of dynamic softening processes in Ti-50.8 at.% Ni aging alloy are established: dynamic recovery in the range 100–300 °C; dynamic polygonization in the range 300–600 °C, dynamic recrystallization – above 600 °C. The inhibition of the dynamic softening processes is defined by the development of dynamic strain aging – the precipitation of the dispersed Ti_3Ni_4 phase particles.

Shape Memory Alloys – SMA 2018   Materials Research Forum LLC
Materials Research Proceedings 9 (2018) 38-42   doi: http://dx.doi.org/10.21741/9781644900017-8

2. The highest recovery characteristics (9 % without PDA and 10–12 % after PDA) were obtained after deformation in the temperature range 400–600 °C. The use of different deformation modes makes it possible to change the ratio of the fractions of the total recoverable strain components, which can be used in the design of devices operating exclusively on the effects of superelasticity or shape memory effects.

3. The use of post-deformation annealing leads to an increase in recovery characteristics under all deformation modes due to the development of dispersion hardening by the precipitation of Ti_3Ni_4 phase particles and is expedient from a technological point of view, since it can be carried out at the stage of the final product manufacturing.

Acknowledgements

The present work was carried out with financial support from the Ministry of Science and Education of Russian Federation (State Task No.11.1495.2017/4.6) and RFBR (research project № 18-08-01193 A).

References

[1] S.D. Prokoshkin, V. Brailovskii, A.V. Korotitskiy, K.E. Inaekyan, A.M. Glezer, Specific features of the formation of the microstructure of titanium nickelide upon thermomechanical treatment including cold plastic deformation to degrees from moderate to severe, Phys. Metals. Metallogr., 110 (2010) 289-303. https://doi.org/10.1134/S0031918X10090127

[2] R.Z. Valiev, I.V. Aleksandrov, Nanostructural Materials Obtained by Severe Plastic Deformation, Integratsiya, Moscow, 2000.

[3] I.Yu. Khmelevskaya, R.D. Karelin, S.D. Prokoshkin, V.A. Andreev, V.S. Yusupov, M.M. Perkas, V.V. Prosvirnin, A.E. Shelest, V.S. Komarov, Effect of the quasi-continuous equal-channel angular pressing on the structure and functional properties of Ti-Ni-based shape-memory alloys, Phys. Metals Metallogr., 118 (2017), 279-287. https://doi.org/10.1134/S0031918X17030073

[4] S. Prokoshkin, I. Khmelevskaya, V. Andreev, R. Karelin, V. Komarov, A. Kazakbiev, Manufacturing of Long-Length Rods of Ultrafine-Grained Ti-Ni Shape Memory Alloys, Mater. Sci. Forum, 918 (2018) 71-76. https://doi.org/10.4028/www.scientific.net/MSF.918.71

[5] I. Khmelevskaya, V. Komarov, R. Kawalla, S. Prokoshkin, G. Korpala, Effect of Biaxial Isothermal Quasi-Continuous Deformation on Structure and Shape Memory Properties of Ti-Ni Alloys, J. Mater. Eng. Perform. 26 (2017) 4011-4019. https://doi.org/10.1007/s11665-017-2841-1

[6] I. Khmelevskaya, V. Komarov, R. Kawalla, S. Prokoshkin, G. Korpala, Features of Ti-Ni alloy structure formation under multi-axial quasi-continuous deformation and post-deformation annealing, Mater. Today: Proc., 4 (2017) 4830-4835. https://doi.org/10.1016/j.matpr.2017.04.079

[7] I.Yu. Khmelevskaya, S.D. Prokoshkin, I.B. Trubitsyna, M.N. Belousov, S.V. Dobatkin, E.V. Tatyanin, A.V. Korotitskiy, V. Brailovski, V.V. Stolyarov, E.A. Prokofiev, Structure and properties of Ti-Ni-based alloys after equal-channel angular pressing and high-pressure torsion, Mater. Sci. Eng. A, 481-482 (2008) 119-122. https://doi.org/10.1016/j.msea.2007.02.157

[8] M.S. Shakeri, J. Khalil-Allafi, V. Abbasi-Chianeh, Arash Ghabchi, The influence of Ni_4Ti_3 precipitates orientation on two-way shape memory effect in a Ni-rich NiTi alloy, J. All. Compd. 485 (2009) 320-323. https://doi.org/10.1016/j.jallcom.2009.05.084

[9] K.A. Polyakova-Vachiyan, E.P. Ryklina, S.D. Prokoshkin, S.M. Dubinskii, Dependence of the functional characteristics of thermomechanically processed titanium nickelide on the size of the structural elements of austenite, Phys. Metals Metallogr. 117 (2016) 817-827. https://doi.org/10.1134/S0031918X16080123

[10] V.S. Komarov, Flow Curves, structure and properties of bulk Ti-Ni alloys, deformed at isothermal conditions, Dissertation (Eng.), NUST MISiS, Moscow, 2018.

Shape Memory Alloys – SMA 2018
Materials Research Proceedings **9** (2018) 43-47

Materials Research Forum LLC
doi: http://dx.doi.org/10.21741/9781644900017-9

Production and Study of the Structure of Novel Superelastic Ti-Zr-Based Alloy

Anton S. Konopatsky[1,a*], Egor M. Barashenkov[1,b], Alibek M. Kazakbiev[1,c], Sergey D. Prokoshkin[1,d]

[1]National University of Science and Technology "MISiS", 4 Leninskiy Prospect, Moscow 119049, Russia

[a]konopatskiy@misis.ru, [b]egor.barashenkov@mail.ru, [c]kazakbiev@yandex.ru, [d]prokoshkin@tmo.misis.ru

*corresponding author

Keywords: Titanium Alloys, Vacuum Arc Remelting, Superelasticity

Abstract. Ti-18Zr-14Nb-4Ta alloy was produced by vacuum arc remelting method. Microstructure and substructure of the alloy in different states were examined. Impurities concentration of the alloy was measured in as-cast state and after homogenizing annealing. Homogeneity of the alloy was studied by scanning electron microscopy (SEM). X-ray diffraction (XRD) study of the alloy after thermomechanical treatment (TMT) revealed high stability of the parent high-temperature β-phase.

Introduction

Shape memory alloys (SMA) were under spotlight of scientific community for many years. For quite a long time TiNi SMA was deservedly considered as the most promising material demonstrating both pronounced shape memory effect (SME) and superelasticity (SE) [1-4]. SME of TiNi is successfully implied in different medical applications such as veins clips [5]. On the other hand development of the material for bone tissue replacement demands not only enhanced and stable SE but also requires using only biocompatible components in alloy's composition. Such desire for safer material led to greater attention to different Ni-free Ti-based SMA [6-9]. Among many different alloying components for Ti-based SMA the following could be outlined: Nb, Zr and Ta. Ternary and quaternary alloys of Ti-Nb-Zr-Ta system were intensively studied for the last 20 years [10-13]. At the start they demonstrated limited SE with up to ~3 % [14] of recovery strain which was considerably lower compared to TiNi. Recently new chemical compositions of the given system such as Ti-18Zr-14Nb (here and after chemical composition is given in at.%, unless otherwise mentioned) were found and significantly higher functional properties were proclaimed [15].

Manufacturing of an alloy composed of these components (Ti, Nb, Zr, Ta) can be quite a difficult task. The bigger the gap between physical properties of the components the harder it is to obtain certain level of alloy's quality. In this case difficulties are connected to the differences between components' melting points. For that reason thorough investigation of the manufacturing process, alloy's microstructure features and chemical composition is required for development of a new systems. In our current study we focus on Ti-18Zr-14Nb-4Ta alloy's manufacturing, heat treatment, thermomechanical treatment (TMT), chemical composition and microstructure study. Introduction of the 4th component (Ta) to Ti-18Zr-14Nb system in a frame of the current study persuades certain goal. Even small concentration of Ta potentially allows to control embrittling ω-phase formation [16]. So investigation of the manufacturing peculiarities for a quaternary alloy appears to be more comprehensive task.

Published under license by Materials Research Forum LLC.

Shape Memory Alloys – SMA 2018   Materials Research Forum LLC
Materials Research Proceedings 9 (2018) 43-47   doi: http://dx.doi.org/10.21741/9781644900017-9

Methods and Materials

Vacuum arc remelting (VAR) furnace with inconsumable W electrode and water cooled Cu mold was used for alloy manufacturing. The mass of the ingot was ~40 g. The ingot was obtained during five consequent remelts with the turning of the ingot upside down after each remelt in order to obtain higher level of homogeneity. Ag protective atmosphere and preparatory remelting of the getter (iodide titanium) were used in order to decrease impurities content.

Homogenization annealing (HA) at 950 °C 60 min in Ag atmosphere with the following quenching in water was implied as the first heat treatment step. Thermomechanical treatment (TMT) consisted of cold rolling with $e = 0.3$ and post deformational annealing at 600 °C 30 min in Ag atmosphere with water quenching was used in order to investigate if the given chemical composition insures SE of the alloy.

Microstructure investigation was conducted by optical microscopy (Versamet-2 Union Carl Zeiss optics) and XRD (Shimadzu XRD-6000) methods. Chemical composition and the uniformity of components distribution were studied by scanning electron microscopy (SEM) with EDX (JEOL5600). Impurities content was measured by high-temperature gas extraction technique.

Results

Typical microstructure of the alloy in different states is given in Fig. 1 (a).

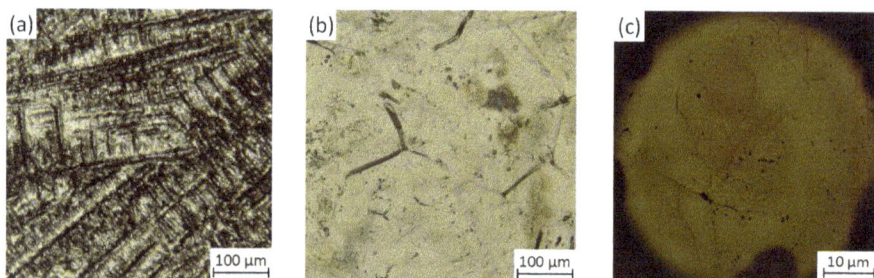

Figure 1. Microstructure of the Ti-18Zr-14Nb-4Ta alloy: (a) as-cast; (b) after homogenization annealing; (c) subgrain structure after homogenizing annealing.

It can be seen from the Fig. 1 that two types of microstructure appear in different states of the alloy: as-cast state is represented by dendritic microstructure (Fig. 1(a)) and after homogenization annealing equiaxed grains could be observed. At closer examination at higher magnifications subgrains structure could also be seen (Fig. 1 (c)).

Electron image of the analyzed surface, integral specter and the corresponding elemental maps are given in Fig. 2.

Fig. 2 (a) depicts the surface of the analyzed area. Fig. 2(b) contains obtained chemical composition of the alloy. Elemental maps (Fig. 2(c)) demonstrate that all components of the alloy are uniformly distributed.

Results of the XRD analysis are given in Fig. 3.

It can be seen from Fig. 3 that phase composition of the alloy is represented exclusively by high-temperature β-phase.

Shape Memory Alloys – SMA 2018
Materials Research Proceedings **9** (2018) 43-47

Materials Research Forum LLC
doi: http://dx.doi.org/10.21741/9781644900017-9

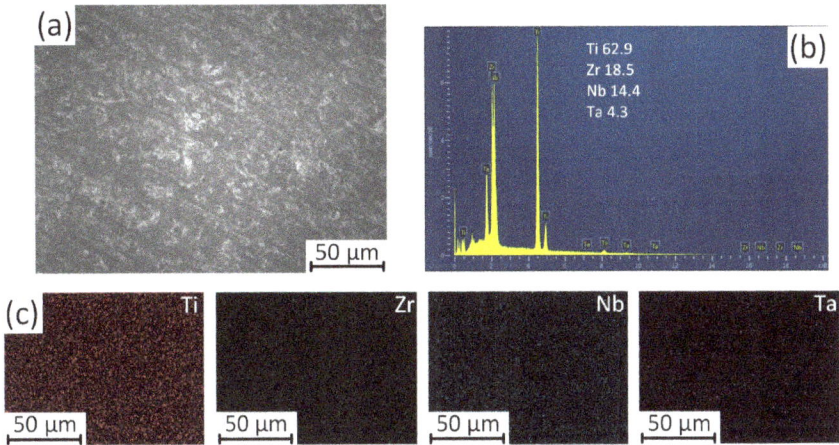

Figure 2. SEM and EDX results: (a) electron image of the analyzed area; (b) integral specter; (c) corresponding elemental maps.

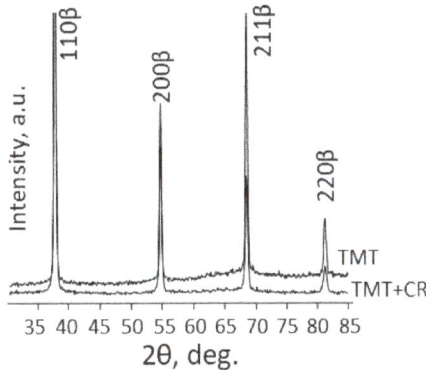

Figure 3. XRD spectra of the Ti-18Zr-14Nb-4Ta alloy after different TMT schemes.

Discussion

In as-cast state alloy possesses dendritic microstructure (Fig. 1(a)). In order to obtain less defective and more equilibrium structure homogenizing annealing was conducted. It can be seen from Fig. 1(b) that after HA much more equilibrium structure of the alloy was obtained. It means that HA parameters allowed to activate diffusion processes needed to remove dendritic segregations. The mean size of the equiaxed grain was about 250 μm. Closer inspection of the microstructure reveals subgrains substructure as well. According to Fig. 1(c) the average size of a subgrain was about 8 μm.

While obtaining less defective and more homogeneous structure is an important task one should take into account that impurities concentration (such as O, N, C, H) significantly affects not only mechanical properties of the titanium alloys but the temperature of the martensitic transformation as well. In order to control the impurities level of the alloy in different states analysis by high-temperature gas extraction technique was conducted. Results are given in Table 1.

Table 1 Impurities content in Ti-18Zr-14Nb-4Ta alloy, at.%.

Element	O	N	C	H
As-cast state	0.0809±0.0002	< 0.001	0.031±0.001	< 0.001
After HA	0.129±0.006	< 0.001	0.037±0.001	< 0.001

Table 1 demonstrates that in as-cast state impurities content is quite low. After HA concentration of the oxygen and carbon increases but insignificantly. Given results confirm that conducted HA allowed to obtain more equilibrium microstructure of the alloy without significant increase of the impurities content.

Obtained elemental maps confirm that five consequent remelts in VAR furnace is enough to obtain high homogeneity of all alloy's components regardless to the considerable difference between their melting points. Good correspondence between nominal and actual chemical composition can be seen from Fig. 2(b).

XRD study revealed that after TMT only high-temperature β-phase can be found on the specter. Crystal lattice period of the observed phase was 3.3495 Å. It is known that small plastic deformation can be effectively used in order to induce martensitic transformation and retain some α''-phase at room temperature [16]. But according to Fig. 3 additional cold rolling (TMT+CR specter) had no effect on the phase composition of the alloy which indicates that the total amount of the β-stabilizing elements was too high and hence the temperature of the martensitic transformation was considerably lower than the room temperature. It means that the given combination of the chemical composition and the TMT scheme is not beneficial for SE of the alloy and decreasing of the β-stabilizing elements could be recommended.

Summary

In the frame of the current study Ti-18Zr-14Nb-4Ta alloy was manufactured. Microstructure of the alloy in as-cast state and after HA was examined. It was shown that homogenizing annealing can be effectively used in order to obtain less defective structure of the ingot without significant increase of the impurities concentration. SEM study demonstrated high homogeneity of the distribution of all alloy's components. XRD results demonstrated that the given chemical composition and TMT scheme lead to formation of the high-temperature β-phase without any trace of the α''-phase even after additional plastic deformation (5–7 %).

Acknowledgement

The present work has been carried out under financial support of the RFBR (grant 18-33-00418 mol_a).

References

[1] S. Miyazaki, K. Otsuka, C.M. Wayman, The Shape Memory Mechanism Associated With the Martensitic-Transformation in Ti-Ni Alloys .1. Self-Accommodation, Acta Metall. 37 (1989) 1873-1884. https://doi.org/10.1016/0001-6160(89)90072-2

[2] C.Y. Nien, H.K. Wang, C.H. Chen, S. Ii, S.K. Wu, C.H. Hsueh, Superelasticity of TiNi-based shape memory alloys at micro/nanoscale, J. Mater. Res. 29 (2014) 2717-2726. https://doi.org/10.1557/jmr.2014.322

[3] S. Prokoshkin, V. Brailovski, K. Inaekyan, V. Demers, A. Kreitcberg, Nanostructured Ti–Ni Shape Memory Alloys Produced by Thermomechanical Processing, Shape Mem. Superelasticity

Shape Memory Alloys – SMA 2018
Materials Research Proceedings 9 (2018) 43-47

Materials Research Forum LLC
doi: http://dx.doi.org/10.21741/9781644900017-9

1 (2015) 191-203. https://doi.org/10.1007/s40830-015-0026-z

[4] S. Prokoshkin, V. Brailovski, S. Dubinskiy, K. Inaekyan, A. Kreitcberg, Gradation of Nanostructures in Cold-Rolled and Annealed Ti--Ni Shape Memory Alloys, Shape Mem. Superelasticity 2 (2016) 12-17. https://doi.org/10.1007/s40830-016-0056-1

[5] E.P. Ryklina, I.Y. Khmelevskaya, S.D. Prokoshkin, R.V. Ipatkin, V.Y. Turilina, K.E. Inaekyan, The nickel-titanium device with SME for emergency interruption of blood flow, Mater. Sci. Eng. A 378 (2004) 519-522. https://doi.org/10.1016/j.msea.2003.12.050

[6] Y.Q. Ma, S.Y. Yang, W.J. Jin, Y.N. Wang, C.P. Wang, X.J. Liu, Microstructure, mechanical and shape memory properties of Ti-55Ta-xSi biomedical alloys, Trans. Nonferrous Met. Soc. 21 (2011) 287-291. https://doi.org/10.1016/S1003-6326(11)60711-5

[7] J. Fu, H.Y. Kim, S. Miyazaki, Effect of annealing temperature on microstructure and superelastic properties of a Ti-18Zr-4.5Nb-3Sn-2Mo alloy, J. Mech. Behav. Biomed. Mater. 65 (2017) 716-723. https://doi.org/10.1016/j.jmbbm.2016.09.036

[8] H.J. Rack, E. Program, Martensitic transformations in Ti-(16–26 at%) Nb alloys, J. Mater. Sci. 31 (1996) 4267-4276. https://doi.org/10.1007/BF00356449

[9] V. Brailovski, S.D. Prokoshkin, M. Gauthier, K. Inaekyan, S. Dubinskiy, M.I. Petrzhik, M. Filonov, Bulk and porous metastable beta Ti-Nb-Zr(Ta) alloys for biomedical applications, Mater. Sci. Eng. C 31 (2011) 643-657. https://doi.org/10.1016/j.msec.2010.12.008

[10] L. W. Ma, C. Y. Chung, Y. X. Tong and Y. F. Zheng, Properties of porous TiNbZr shape memory alloy fabricated by mechanical alloying and hot isostatic pressing, J. Mater. Eng. Perform. 20 (2011) 783-786. https://doi.org/10.1007/s11665-011-9913-4

[11] M. Tahara, N. Okano, T. Inamura and H. Hosoda, Plastic deformation behaviour of single-crystalline martensite of Ti-Nb shape memory alloy, Sci. Rep. 7 (2017) 1-11. https://doi.org/10.1038/s41598-016-0028-x

[12] H.Y. Kim, S. Miyazaki, Martensitic Transformation and Superelastic Properties of Ti-Nb Base Alloys, Mater. Trans. 56 (2015) 625-634. https://doi.org/10.2320/matertrans.M2014454

[13] N. Sakaguchi, M. Niinomi, T. Akahori, J. Takeda, H. Toda, Effect of Ta content on mechanical properties of Ti-30Nb-XTa-5Zr, Mater. Sci. Eng. C, 25 (2005) 370-376. https://doi.org/10.1016/j.msec.2005.04.003

[14] H.Y. Kim, T. Sasaki, J.I. Kim, T. Inamura, H. Hosoda, S. Miyazaki, Texture and shape memory behavior of Ti-22Nb-6Ta alloy, Acta Mater. 54 (2006) 423-433. https://doi.org/10.1016/j.actamat.2005.09.014

[15] H.Y. Kim, J. Fu, H. Tobe, J. Il Kim, S. Miyazaki, Crystal Structure, Transformation Strain, and Superelastic Property of Ti–Nb–Zr and Ti–Nb–Ta Alloys, Shape Mem. Superelasticity 1 (2015) 107-116. https://doi.org/10.1007/s40830-015-0022-3

[16] A.S. Konopatsky, S.M. Dubinskiy, Yu.S. Zhukova, V. Sheremetyev, V. Brailovski, S.D. Prokoshkin, M. R. Filonov, Ternary Ti-Zr-Nb and quaternary Ti-Zr-Nb-Ta shape memory alloys for biomedical applications: Structural features and cyclic mechanical properties, Mater. Sci. Eng. A 702 (2017) 301-311. https://doi.org/10.1016/j.msea.2017.07.046

Shape Memory Alloys – SMA 2018
Materials Research Proceedings 9 (2018) 48-52

Materials Research Forum LLC
doi: http://dx.doi.org/10.21741/9781644900017-10

Influence of Stress-induced Martensite Ageing on the Shape Memory Effects in As-grown and Quenched [011]-oriented Single Crystals of Ni$_{49}$Fe$_{18}$Ga$_{27}$Co$_6$ Alloy

Aida B. Tokhmetova[1,a*], Natalia G. Larchenkova[1,b], Elena Yu. Panchenko[1,c], Ekaterina E. Timofeeva[1,d], Nikita Yu. Surikov[1,e], Yury I. Chumlyakov[1,f]

[1]National Research Tomsk State University, 36 Lenin Ave, Tomsk 634050, Russia

[a]Aida-tx@mail.ru, [b]Vetnat23@gmail.com, [c]Panchenko@mail.tsu.ru, [d]Katie@sibmail.com, [e]Jet_n@mail.ru, [f]Chum@phys.tsu.ru

*corresponding author

Keywords: Single Crystals, Stress-Induced Martensite Aging, Shape Memory Effect

Abstract. The influence of the stress-induced martensite aging on the two-way shape memory effect on Ni$_{49}$Fe$_{18}$Ga$_{27}$Co$_6$ (at.%) single crystals was studied. It has been shown, that the optimal regime of martensite aging for inducing the two-way shape memory effect is aging at $T = 373$ K, 1 h under a compressive stress 300 MPa applied along the [011]$_A$-direction. After this ageing the two-way shape memory effect was obtained along [001]$_A$-orientation in quenched B2-crystals with reversible strain 7.0 %, and in as-grown (L2$_1$ + γ)-crystals with strain 4.5 %.

Introduction

Ni$_{49}$Fe$_{18}$Ga$_{27}$Co$_6$ single crystals are known due to the thermoelastic martensitic transformations (MT) and related effects – superelasticity (SE), one-way and two-way shape memory effects (SME and TWSME) [1,2]. It is known, that there is no need to apply a stress for the reorientation of the self-accommodating structure of a thermal-induced martensite and growth of an oriented martensite in alloys with TWSME. Therefore, a sample with TWSME reverses its form during heating/cooling cycles without external stresses in contrast to alloys with one-way SME. It is possible to induce TWSME in alloys by various trainings under stress. For example, in the [001]$_A$-oriented Ni$_{49}$Fe$_{18}$Ga$_{27}$Co$_6$ single crystals the maximum TWSME strain up to $\varepsilon = 5.5$ % was obtained after a stepped thermomechanical treatment (annealing at $T = 1373$ K quenching + aging at $T = 673$ K under external stress 100 MPa) and mechanical training (100 loading/unloading cycles at $T = 295$ K) [3]. However, with an increase in cooling/heating cycles up to 100 cycles the reversible TWSME strain degrades by 45 % [3]. Based on the above, it is necessary to search for new thermomechanical treatment for obtainment the TWSME with a maximum reversible strain and a high cyclic stability of functional properties. It is known [4] that the stress-induced martensite aging (SIM-ageing) is effective thermomechanical treatment, which is still not studied in Ni$_{49}$Fe$_{18}$Ga$_{27}$Co$_6$ single crystals. Therefore, the aim of the work is to investigate the effect of SIM-ageing on TWSME in as-grown and quenched Ni$_{49}$Fe$_{18}$Ga$_{27}$Co$_6$ single crystals, oriented along the [011] direction.

Materials and method

Ni$_{49}$Fe$_{18}$Ga$_{27}$Co$_6$ single crystals were grown by the Bridgman method in an inert gas atmosphere. Samples for compression tests were electro-discharge machined in the form of a rectangular parallelepiped with sizes of 3×3×6 mm^3. SIM-ageing was conducted along the [011]-direction. The maximum theoretically calculated transformation strain including the strain associated with a formation of the CVP of L1$_0$-martensite ε_{CVP} and its subsequent detwinning ε_{detw} is $\varepsilon_{tr} = \varepsilon_{CVP} + \varepsilon_{detw} = 6.2$ %. The efficiency of such aging in a detwinned L1$_0$-martensite under stress applied along [011]$_A$||[100]$_M$ direction is shown in [5]. The investigations of TWSME and SE

Published under license by Materials Research Forum LLC.

Shape Memory Alloys – SMA 2018
Materials Research Proceedings **9** (2018) 48-52

Materials Research Forum LLC
doi: http://dx.doi.org/10.21741/9781644900017-10

were carried out on single crystals of two types: an initial as-grown crystals – (L2$_1$+γ)-crystals, and after the high-temperature annealing at $T = 1448$ K, 1 h followed by quenching in water – B2-crystals. The optical metallography was studied on the digital measuring microscope Keyence VHX-2000. The SE experiments and the SIM-ageing were carried out on the mechanical testing machine Instron 5969 with a strain rate $\varepsilon = 2 \times 10^{-3}$ sec^{-1}. Mechanical tests for the observation of SME and TWSME were carried out using the special testing frame, which strain measurement error does not exceed 0.3 %.

Experimental results and discussion
The optical metallography investigations and X-ray diffraction analysis have shown that the as-grown Ni$_{49}$Fe$_{18}$Ga$_{27}$Co$_6$ single crystals is heterophase and consist of L2$_1$-structured matrix with γ-phase particles (Fig. 1 a). The γ-phase particles have a fcc lattice and do not undergo MT [6]. A scanning electron microscope investigation showed, that the γ-phase particles are enriched in Fe, Co, and depleted in Ga (Table 1, Fig. 1 b), which agrees with the data in [7]. The transition from the L2$_1$-austenite to L1$_0$-martensite occurs through the layered modulated 14M-martensite. The annealing at a temperature higher than the order-disorder transition temperature ~ 975 K [8] and the subsequent quenching into water lead, first, to the change in the high-temperature structure from L2$_1$ to the B2, while the γ-phase particles are dissolved (Fig. 1 b). Second, after the annealing the MT sequence becomes B2-L1$_0$ MT.

Figure 1. Optical metallography of the Ni$_{49}$Fe$_{18}$Ga$_{27}$Co$_6$ single crystals surface: (a) (L2$_1$ + γ)-crystals; (b) B2-crystals.

Table 1. Chemical composition of the matrix and particles in as-grown Ni$_{49}$Fe$_{18}$Ga$_{27}$Co$_6$ single crystals.

Element	Ni	Fe	Ga	Co
at.% particles	49	25	18	8
at.% matrix	49	18	27	6

The characteristic temperatures of start and end of forward and reverse MT in (L2$_1$ + γ)- and B2-crystals upon stress-free cooling/heating (Table 2) are determined from the temperature dependence of the resistivity $\rho(T)$.

Table 2. MT characteristic temperatures of Ni$_{49}$Fe$_{18}$Ga$_{27}$Co$_6$ (L2$_1$ + γ)- and B2-crystals.

Crystals	Characteristic temperatures					
	M_s, (± 2) K	M_f, (± 2) K	A_s, (± 2) K	A_f, (± 2) K	Δ_1, (± 2) K	Δ_2, (± 2) K
(L2$_1$+γ)	260	256	265	270	4	5
B2	263	239	258	277	24	19

The SE was investigated at $T > A_f$ in a wide temperature range from $T = 297$ K to $T = 423–448$ K in (L2$_1$ + γ)- and B2-crystals (Fig. 2). MT in both crystals at $T = 297$ K occurs with a low critical stress level of martensite formation σ_{cr}. In contrast to B2-crystals, in (L2$_1$ + γ)-crystals at $T = 297$ K the L1$_0$-martensite is stabilized at the loading/unloading cycles, and the

Shape Memory Alloys – SMA 2018   Materials Research Forum LLC
Materials Research Proceedings **9** (2018) 48-52   doi: http://dx.doi.org/10.21741/9781644900017-10

given strain returns only after heating above the $T = A_f + 30$ K. Such an effect of martensite stabilization, that leads to rise of A_s and A_f temperatures and a burst-like behavior during the reverse transformation were observed earlier in NiFeGaCo single crystals [9]. With increase the test temperature up to 373 K, according to the Clapeyron-Clausius equation [10], in both crystals the stress σ_{cr} grows up to 253 MPa, and also the SE with completely reversible strain is observed. At $T = 297$ K and $T = 373$ K the SE curves show two stages (Fig. 2). With a further increase in temperature until to $T = 448$ K, the SE curves demonstrate a one-stage transformation. In both crystals, σ_{cr} raises up to 350 MPa, and irreversible strain up to 0.7 % is observed.

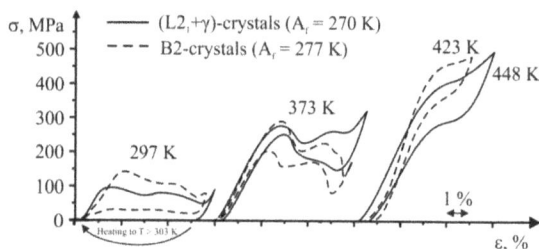

Figure 2. Stress-strain response in the superelasticity temperature range in [011]-oriented $Ni_{49}Fe_{18}Ga_{27}Co_6$ single crystals.

Following the foresaid the SIM-aging was carried out at $T = 373$ K, when a complete reversible strain is observed in both $(L2_1 + \gamma)$- and B2-crystals. For SIM-aging the sample was loaded at $T = 373$ K until it transforms from the $L2_1$/B2 structure into the $L1_0$-martensite, and then after 1 hour of holding with a strain $\varepsilon_{set} = 6.0$–7.0 % $> \varepsilon_{CVP} = 3.2$ % were unloaded (Fig. 3). In B2-crystals at $\varepsilon_{set} = 6.0$ % the corresponding aging stress $\sigma_{ag1} = 200$ MPa is less, than σ_{cr1}, by 47 MPa and at $\varepsilon_{set} = 7.0$ %, σ_{ag2} increases to 300 MPa. In the $(L2_1 + \gamma)$-crystals, even at $\varepsilon_{set} = 6.0$ %, σ_{ag3} is equals to 300 MPa. Thereby, two SIM-ageing regimes were developed:
1. $T = 373$ K, 1 h under stress $\sigma_{ag} = 200$ MPa – regime I;
2. $T = 373$ K, 1 h under stress $\sigma_{ag} = 300$ MPa – regime II.

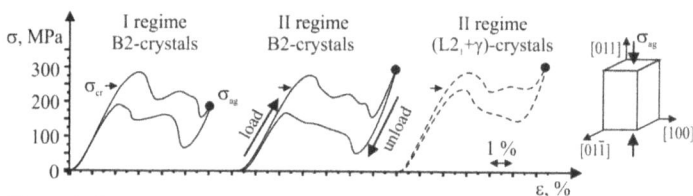

Figure 3. Stress-strain response during the SIM-ageing along [011]-direction of $Ni_{49}Fe_{18}Ga_{27}Co_6$ single crystals.

SIM-ageing results in $L1_0$-martensite stabilization, which is accompanied with the compression along $[011]_A$-direction and tensile along the perpendicular direction at MT (Fig. 3). Thus, in the subsequent cooling/heating cycles the SIM-aged crystals demonstrate the TWSME. After regime I in B2-crystals, when $\sigma_{cr1} > \sigma_{ag1}$, the TWSME strain is only 0.5 %, which indicates the ineffectiveness of this regime (Fig. 4). After regime II in B2-crystals at $\sigma_{cr1} < \sigma_{ag1}$, the TWSME strain is 7.0 %, and thermal hysteresis is $\Delta T = 57$ K (Fig. 5). In the $(L2_1 + \gamma)$-crystals, after regime II the TWSME with reversible strain up to $\varepsilon = 4.3$ % and a narrow stress hysteresis $\Delta T = 17$ K is observed. The main mechanism of martensite stabilization after SIM-aging, like in [4,11] is, firstly, a chemical stabilization. The symmetry-conforming short-range order theory [4] indicates that the SIM-ageing leads to a change in the probability of the distribution of atoms of different kinds in sublattices, and the short-range order of point defects [4]. Secondly, during the SIM-aging along the $[011]_A$-direction, a mechanical stabilization of $L1_0$-martensite is possible due to the detwinning and reorientation of the habit plane. So in B2-crystals TWSME was accompanied by a wide hysteresis and reversible strain $\varepsilon = 7$ % more than the theoretically strain of formation of twinned $L1_0$-martensite $\varepsilon_{CVP}^{L10} = 6.2$ % in tension. In contrast to the B2-crystals, the $(L2_1 + \gamma)$-crystals demonstrate the TWSME $\varepsilon_{TWSME} = 4.5$ % $< \varepsilon_{CVP}^{14M-L10} = 6.2$ % with a narrow thermal hysteresis, because MT passes through the twinned 14M-martensite and provides a low frictional force for the motion of interphase boundaries.

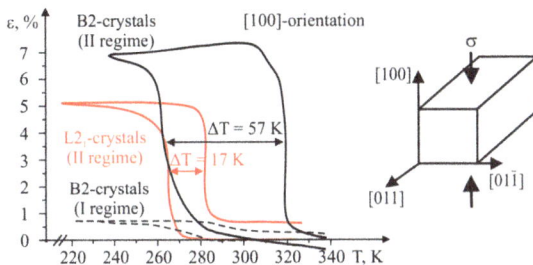

Figure 4. TWSME response along [100]-direction in SIM-aged $Ni_{49}Fe_{18}Ga_{27}Co_6$ single crystals.

Summary

1. An effective SIM-ageing regime for inducing the TWSME ($T = 373$ K, 1 h under stress $\sigma_{ag} = 300$ MPa along the [011]-orientation) was determined in as-grown $(L2_1 + \gamma)$-crystals and quenched B2-crystals of $Ni_{49}Fe_{18}Ga_{27}Co_6$ alloy.

2. It has been shown, that SIM-ageing results in inducing of TWSME along the $[100]_A$-direction with a reversible strain of $\varepsilon = 4.3$ % in $(L2_1 + \gamma)$-crystals, and $\varepsilon = 7.0$ % in B2-crystals.

3. The thermal hysteresis strongly depends on the microstructure of the austenite and the MT sequence. In B2-crystals the thermal hysteresis $\Delta T = 57$ K at B2-$L1_0$ MT is three times larger than in hysteresis $\Delta T = 17$ K $(L2_1 + \gamma)$-crystals, which undergo MT in cooling/heating cycles into a layered modulated 14M-martensite.

Acknowledgements

This work was supported by the Russian Science Foundation (grant No. 16-19-10250).

References

[1] R.F. Hamilton, H. Sehitoglu, C. Efstathiou, H.J. Maier, Inter-martensitic transitions in Ni–Fe–Ga single crystals, Acta Mater. 55 (2007) 4867-4876. https://doi.org/10.1016/j.actamat.2007.05.003

[2] E.E. Timofeeva, E.Yu. Panchenko, Yu.I. Chumlyakov, N.G. Vetoshkina, H.J. Maier, One-way and two-way shape memory effect in ferromagnetic NiFeGaCo single crystals, Mater. Sci. Eng. A. 378 (2004) 403-408. https://doi.org/10.1016/j.msea.2003.10.366

[3] E. Yu. Panchenko, E.E. Timofeeva, N.G. Larchenkova, Y.I. Chumlyakov, A.I. Tagiltsev, H.J. Maier, G. Gerstein, Two-way shape memory effect under multi-cycles in [001]-oriented $Ni_{49}Fe_{18}Ga_{27}Co_6$ single crystal, Mater. Sci. Eng. A. 706 (2017) 95-103. https://doi.org/10.1016/j.msea.2017.08.108

Shape Memory Alloys – SMA 2018   Materials Research Forum LLC
Materials Research Proceedings **9** (2018) 48-52   doi: http://dx.doi.org/10.21741/9781644900017-10

[4] K. Otsuka, X. Ren, Mechanism of martensite aging effects and new aspects, Mater. Sci. Eng. A. 312 (2001) 207-218. https://doi.org/10.1016/S0921-5093(00)01877-3

[5] E. Panchenko, A. Eftifeeva, Y. Chumlyakov, G. Gerstein, H.J. Maier, Two-way shape memory effect and thermal cycling stability in $Co_{35}Ni_{35}Al_{30}$ single crystals by low-temperature martensite ageing, Scripta Mater. 150 (2018) 18-21. https://doi.org/10.1016/j.scriptamat.2018.02.013

[6] N. Vetoshkina, E. Panchenko, E. Timofeeva, Yu. Chumlyakov, N. Surikov, K. Osipovich, H. Maier, Effects of ageing on the cyclic stability of superelasticity in [001]-oriented $Ni_{49}Fe_{18}Ga_{27}Co_6$ single crystals in compression, Mater. Today: Proc. (2017) 4797-4801.

[7] Y. Imano, T. Omori, K. Oikawa, Y. Sutou, R. Kainuma, K. Ishida, Martensitic and magnetic transformations of Ni–Ga–Fe–Co ferromagnetic shape memory alloys, Mater. Sci. Eng. A. 438–440 (2006) 970-973. https://doi.org/10.1016/j.msea.2006.02.080

[8] T. Omori, N. Kamiya, Y. Sutou, K. Oikawa, R. Kainuma, K. Ishida, Phase transformations in Ni–Ga–Fe ferromagnetic shape memory alloys, Mater. Sci. Eng. A. 378 (2004) 403-408. https://doi.org/10.1016/j.msea.2003.10.366

[9] Dewei Zhao, Fei Xiao, Zhihua Nie, Daoyong Cong, Wen Sun, Jian Liu, Burst-like superelasticity and elastocaloric effect in [011] oriented $Ni_{50}Fe_{19}Ga_{27}Co_4$ single crystals, Scripta Mater. 149 (2018) 6-10. https://doi.org/10.1016/j.scriptamat.2018.01.029

[10] Y.I. Chumlyakov, E.Y. Panchenko, A.V. Ovsyannikov, S.A. Chusov, V.A. Kirillov, I. Karaman, H.J. Maier, High-temperature superelasticity and the shape memory effect in [001] Co-Ni-Al single crystals, Phys. Met. Metallogr. 107 (2009) 194-205. https://doi.org/10.1134/S0031918X09020124

[11] S. Kustov, J. Pons, E. Cesari, J. Van Humbeeck, Chemical and mechanical stabilization of martensite, Acta Mater. 52 (2004) 4547-4559. https://doi.org/10.1016/j.actamat.2004.06.012

Shape Memory Alloys – SMA 2018
Materials Research Proceedings 9 (2018) 53-57

Materials Research Forum LLC
doi: http://dx.doi.org/10.21741/9781644900017-11

Solid State Cooling Based on Elastocaloric Effect in Melt Spun Ribbons of the Ti₂NiCu Alloy

Evgeny V. Morozov[1,a*], Sergey Y. Fedotov[1,2,b], Maria S. Bibik[1,c],
Aleksey V. Petrov[1,d], Victor V. Koledov[1,e], Vladimir G. Shavrov[1,f]

[1]Kotelnikov Institute of Radio-engineering and Electronics of RAS, 11-7 Mokhovaya Str.,
Moscow 125009, Russia

[2]Lomonosov Moscow State University, Physical Faculty, 1/2 Leninskie Gory, Moscow 119991,
Russia

[a]evgvmorozov@gmail.com, [b]ser52@list.ru, [c]bybik.m.s@gmail.com, [d]alexvc2003@mail.ru,
[e]victor_koledov@mail.ru, [f]shavrov@cplire.ru

*corresponding author

Keywords: Elastocaloric Effect, Shape Memory, Solid State Cooling, Melt Spun Ribbons, Ti₂NiCu Alloy, Multicaloric Effect, Refrigeration, Intermetallic Alloy, Rapidly Quenched Ribbons, COP, IR-Thermography

Abstract. This paper is devoted to the experimental study of elastocaloric effect (ECE) in rapidly quenched ribbons of the Ti₂NiCu alloy with shape memory effect (SME) under the exposure of periodic tensile force. ECE is measured depending on the relative strain and frequency of the cycles in range from 0.2 to 4 Hz. The maximal measured ECE in the alloy reaches 9.4 K at a mechanical stress of 300 MPa, with the relative strain of the sample being equal to 1 %, and at frequencies from 0.2 to 0.5 Hz at temperature 67 °C. The specific power of the heat exchange of the working body with ECE was estimated. Specific power reaches a maximal value $W = 10$ W/g at frequency $f = 4$ Hz. The prospects of the design of the solid state cooling devices based on ECE in Ti₂NiCu alloy was discussed.

Introduction

Nowadays, a great attention of scientists and engineers all over the world is paid to the development of new functional materials. These materials such as alloys with phase transitions in which the external fields cause a sharp entropy and temperature change are recognized as promising for the future application in the field of alternative energy technologies [1-3]. For obtaining these objectives, elastocaloric, magnetocaloric and electrocaloric etc. effects are studying all over the world. These effects are associated with a change of the temperature under the influence of correspondent external fields: the mechanical, magnetic and electrical fields, respectively. This work is devoted to the experimental study of elastocaloric effect (EEC) under the exposure of tensile force in Ti₂NiCu alloy with shape memory effect. During the last 5 years elastocaloric effect attracts the growing interest [4-5]. It is shown, that ECE in TiNi alloy has a value up to 16 K [2]. However, issues devoted to the ECE dependence on the frequency of cycles have not been studied sufficiently yet.

The purpose of the present work was to study ECE under the exposure of periodic tensile force to obtain the dependence of ECE on frequency of elongation cycles and to discuss the prospects of possible applications of ECE in rapidly quenched ribbons of Ti₂NiCu alloy for development of the alternative energy devices.

Experimental

The material under investigation was manufactured by a single-roller melt-spinning technique from a pre-synthesized Ti50Ni25Cu25 (at.%) alloy. High purity nickel, titanium and copper were

Published under license by Materials Research Forum LLC.

Shape Memory Alloys – SMA 2018   Materials Research Forum LLC
Materials Research Proceedings 9 (2018) 53-57   doi: http://dx.doi.org/10.21741/9781644900017-11

melted six times in an argon arc furnace. The obtained ingots were melted down in quartz crucibles under a purified atmosphere of inert gas and ejected onto the surface of a fast-rotating copper wheel at cooling rates around 106 K/s. Thus, the rapidly quenched $Ti_{50}Ni_{25}Cu_{25}$ alloy was prepared in the form of continuous (10–30 m long) ribbon with a thickness in the range 30–50 μm and a width about 1.5 mm. The shape memory effect in this material is due to thermoelastic martensitic transition of 2d order, which has characteristic transition temperatures of the start and finish of direct and inverse transitions M_s = 42 °C, M_f = 39 °C, A_s = 50 °C, and A_f = 52 °C, respectively [7]. The ECE can be treated as effect which is essantualy is inverse effect compared with SME. So maximal adiabatic effect temperature change under external stress is expected near A_f = 52 °C.

The experimental setup for ECE measurements is shown schematically in Fig. 1. The setup comprises the sample of rapidly quenched alloy Ti2NiCu 1, actuator 2 rigidly fixed on the platform with a stroke length of 22 mm, connected to voltage pulse generator 7. The actuator provides periodic deformation of the sample 1, rigidly fixed at one end. The other end is connected to a spring, which is in turn fixed to actuator. The sample is a rapidly quenched ribbon with length 10 sm, width 1.5 mm and thickness 35 μm. The initial temperature of the sample is controlled by heater 3, which is a dielectric tube wrapped in nichrome wire connected to DC voltage supply 6. The sample surface temperature is measured by Testo-845 pyrometric sensor 5 with a frame frequency of 10 Hz. The pyrometer is focused on the centre of the studied sample to measure the temperature at a certain fixed point. The data was obtained and processed by PC 8, equipped with the L-Card E14-140M ADC. In addition, for the thermographic studies, the Flir SC- 7000 thermal imager was used instead of the pyrometer sensor.

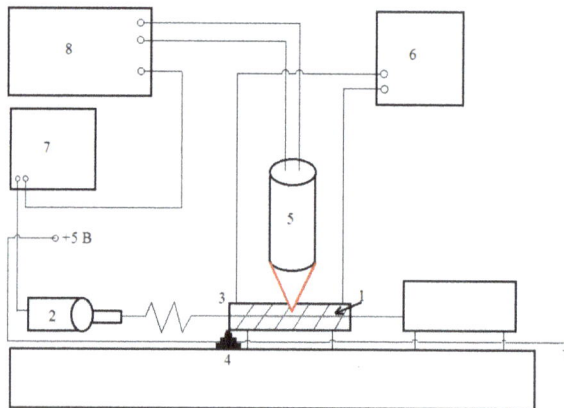

Figure 1. Experimental setup for studying ECE.

The measurements were done in frequency range from 0.2 Hz to 4 Hz. Relative deformation of the sample during the experiments varied up to 1 %.

Results and discussing

Fig.2 shows the example of the temporal dependences of ECE on temperature of sample, under the periodic loads in frequency range from 0.2 to 4 Hz. The maximum measured ECE was 9.4 K at 0.2 Hz and 300 MPa of external stress (see Fig. 2 (c)). Figure 3 (a) shows the ECE as a function of frequency at different loads: 150 MPa, 225 MPa and 300 MPa. Based on these curves, we concluded, that the ECE diminished at the frequency of 2 Hz.

Shape Memory Alloys – SMA 2018   Materials Research Forum LLC
Materials Research Proceedings **9** (2018) 53-57   doi: http://dx.doi.org/10.21741/9781644900017-11

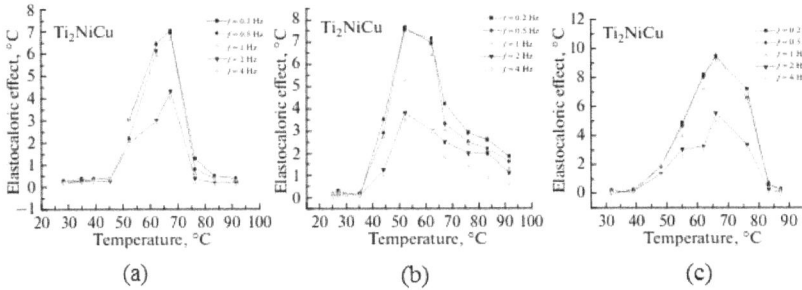

Figure 2. ECE as a function of temperature in Ti$_2$NiCu alloy under the different loads:
a) 150 MPa, b) 225 MPa, c) 300 MPa.

The considerations of possibility of practical use of Ti$_2$NiCu alloy for the solid- state cooling have been also occurred. The specific power of samples was calculated as:

$$W = c \cdot \Delta T \cdot f,\qquad(1)$$

where W – specific power, c – specific heat capacity, ΔT – temperature change during the deformation (ECE), f – frequency of deformation cycles. Maximum specific power was 10 W/g at the load of 300 MPa and frequency of 4 Hz (Fig.3 (b)).

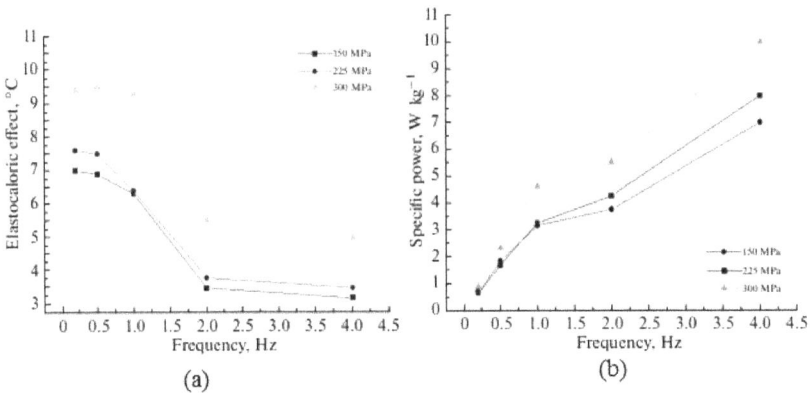

Figure 3. ECE (a) and specific power (b) dependences on the frequency of deformation cycles.

The obtained results show prospects of the applications of the studied functional material with SME and ECE – Ti$_2$NiCu melt spun ribbons, because the value of the specific power is relatively big, compared with existing functional materials. These results of Ti$_2$NiCu alloy are competitive to those of the materials with magnetocaloric effect (MCE) in high magnetic field up to 10 Tesla. Several questions arise in this connection. The most important is the nature of the effect of the drop of the ECE with the rise of the frequency of the cycles. The more comprehensive experimental study is to be conducted in order to clarify the peqularities of the ECE in the frequency range up to 100 Hz. For this purpose, the thermographic studies were continued using the Flir SC-7000 thermal imager with higher space and temporal resolution (see example of thermal images on the Fig. 4).

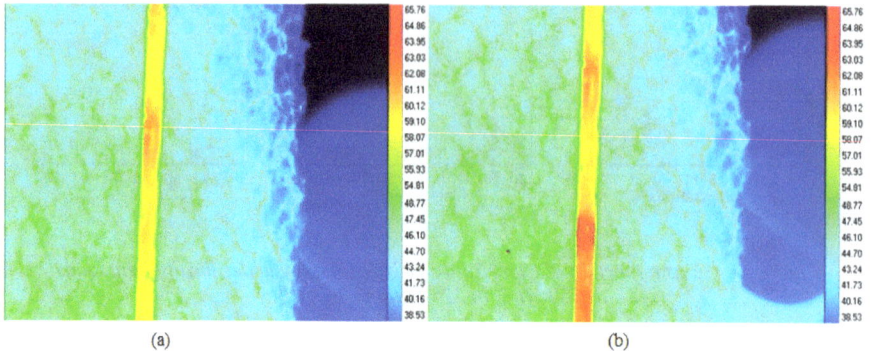

Figure 4. Registration of ECE, using IR-thermography with high space and temporal resolution.

IR-thermography pictures at Fig. 4 show the ECE in fast-quenched ribbons of Ti_2NiCu alloy under the periodic tensile force with different loads: (a) – 225 MPa and (b) – 300 MPa respectively. Both images show, that ribbons are heterogeneous, i.e. phase transition in these samples occurs in different ways. IR-termography gives us a possibility to increase the space and temporal resolution and, moreover, the accuracy of measurements.

Conclusions

We may formulate the results of the present studies as follows.

1) An experimental setup for studying of ECE in ribbons of different materials under the periodic tensile force was created and modified for further high- frequent studying up to 30 Hz.

2) ECE was studied in the range of frequencies from 0.2 Hz to 4 Hz under the different mechanical loads: 150 MPa, 225 MPa and 300 MPa. The maximum measured ECE in Ti_2NiCu alloy is 9.4 K at 0.2 Hz and 300 MPa.

3) At frequency of 2 Hz ECE drops sharply. This effect can be attributed to the kinetic processes during the first order phase transition.

4) Specific power of Ti_2NiCu alloy samples was estimated. The maximal value of specific power is about 10 W/g. This result is competitive to the results in MCE- materials.

5) A new method of registration and studying of ECE in ribbons, based on IR- thermography, was introduced and tested during the experiments. This approach is considered as promising for the further studying of kinetic processes in alloys during the first order phase transition.

Acknowledgments

This work was supported by the Russian Foundation for Basic Research, grant No 18-37-00481.

References

[1] S.A. Nikitin, G. Myalikgulyev, M.P. Annoaorazov, A.L. Tyurin, R.W. Myndyev, S.A. Akopyan, Giant elastocaloric effect in FeRh alloy, Phys. Lett. A 171 (1992) 234-236. https://doi.org/10.1016/0375-9601(92)90432-L

[2] Manosa L., Planes A., Vives E., Bonnot E., Romero R., The use of shape-memory alloys for mechanical refrigeration, Func. Mater. Lett. 2 (2009) 73-78. https://doi.org/10.1142/S1793604709000594

[3] H. Ossmer, C. Chluba, S. Kauffmann-Weiss, E. Quandt, M. Kohl, TiNi-based films for elastocaloric microcooling. Fatigue life and device performance, APL Materials 4 (2016) 064102. https://doi.org/10.1063/1.4948271

[4] Guyomar, D., Li Y., Sebald G., Cottinet P., Ducharne B., Capsal J., Elastocaloric modeling of natural rubber, Appl. Therm. Eng. 57 (2013) 33-38. https://doi.org/10.1016/j.applthermaleng.2013.03.032

[5] A.P. Kamantsev, V.V. Koledov, A.V. Mashirov, E.T. Dilmieva, V.G. Shavrov, J. Cwik, I.S. Tereshina, Direct measurement of magnetocaloric effect in metamagnetic Ni43Mn37.9In12.1Co7 Heusler Alloy, Bull. Russ. Acad. Sci.: Phys. 78 (2014) 936- 938. https://doi.org/10.3103/S106287381409010X

[6], A.V. Shelyakov, N.N. Sitnikov, V.V. Koledov, D.S. Kuchin, A.I. Irzhak, N.Y. Tabachkova, Melt-spun thin ribbons of shape memory TiNiCu alloy for micromechanical applications, International Journal of Smart and Nano Materials, 2 (2011) 68-77. https://doi.org/10.1080/19475411.2011.567305

[7] P. Lega, V. Koledov, A. Orlov, D. Kuchin, A. Frolov, V. Shavrov, V Khovaylo. Composite Materials Based on Shape-Memory Ti2NiCu Alloy for Frontier Micro- and Nanomechanical Applications, Adv. Eng. Mater. 19 (2017) 1700154. https://doi.org/10.1002/adem.201700154

Shape Memory Alloys – SMA 2018
Materials Research Proceedings 9 (2018) 58-62

Materials Research Forum LLC
doi: http://dx.doi.org/10.21741/9781644900017-12

Mechanical Properties of the TiNi and Surface Alloy Formed by Pulsed Electron Beam Treatment

Alexey A. Neiman[1,a*], Regina R. Mukhamedova[1,2,b], Viktor O. Semin[1,2,c]

[1]Institute of Strength Physics and Material Science SB RAS, 634055 Tomsk, Russia

[2]National Research Tomsk State University, 634050 Tomsk, Russia

[a]nasa@ispms.tsc.ru, [b]reginagforce@gmail.com, [c]lpfreedom14@gmail.com

*corresponding author

Keywords: Shape Memory Alloys, Electron Beam Treatment, Superelasticity, Hardness

Abstract. The mechanical properties of the TiNi-based alloy irradiated with microsecond low-energy high-current pulsed electron beam (LEHCPEB) as well as the Ti-Ta-Ni surface alloy were investigated. It was shown that surface modification by LEHCPEB of the TiNi alloy increases the hardness on average by 20 % compared to its initial value. Stress-strain curves as a function of temperature, corresponding to different structure (martensitic and austenitic) states of the TiNi alloy, displayed pseudoelastic behavior with almost completely recovering strain upon unloading.

Introduction

TiNi-based alloys are known to be biomedically applicable shape memory alloy for surgical implants and medical devices owing to their superior biocompatibility and capability for strain recovery on heating to human body temperature [1]. The requirements that must be met by metallic biomaterials include, in addition to biocompatibility, also high corrosion resistance, functional mechanical properties and fatigue resistance. For superelastic TiNi alloy two crucial factors limiting its applications, for example in orthopedics or cardiovascular surgery, are the high content (~ 50 at.%) of toxic Ni and a relatively low fatigue properties.

Surface treatment of metals and alloys by low-energy high-current pulsed electron beam (LEHCPEB) in modes of surface melting follows special demands concerning an enhancement of the physical-mechanical properties, namely, to allows to increase the fatigue strength by more than 20 %, the fatigue life more than tenfold [2]. Previous studies [3] have shown that the LEHCPEB treatment of TiNi alloys leads to the formation of a multiplayer surface structure in the heat-affected zone characterized by high superelasticity characteristics and hardness in the upper surface layer. In turn, the Ti-Ta-based surface alloys fabricated on the TiNi substrate feature high-elasticity, low plasticity, and high hardness. The superelastic materials subjected to cyclic loading in the body tissues must exhibit pseudoelastic behaviour in the temperature interval of direct and reverse martensitic transformations. Surface treatments of the TiNi-based alloys are accompanied by the formation of nonequilibrium microstructures, including amorphous and nanocrystalline ones, high residual stresses, that inevitably leads to the change in the elastic-plastic characteristics of surface layers.

Since superelastic properties are very sensitive to the structure-phase states formed after surface treatments, the aim of the present work is to gain further insights in the influence of the LEHCPEB treatment and formation of the surface alloy on the mechanical (pseudoelastic) properties of the TiNi alloy.

Experimental work

The substrate material was TiNi alloy produced from pure titanium and nickel ($\leq$ 99.99 wt.% both) by six-fold vacuum arc remelting with further vacuum annealing (1073 K, 10^{-3} Pa, 1 h) and

Published under license by Materials Research Forum LLC.

Shape Memory Alloys – SMA 2018   Materials Research Forum LLC
Materials Research Proceedings **9** (2018) 58-62   doi: http://dx.doi.org/10.21741/9781644900017-12

furnace cooling. The chemical composition of the B2 matrix phase was $Ti_{49.5}Ni_{50.5}$ (at. %), and its transformation temperatures, measured by temperature X-ray diffraction method, were $M_s = 290 \pm 1$ K, $M_f = 270 \pm 1$ K, $A_s = 303 \pm 1$ K, and $A_f = 330 \pm 1$ K. Before electron beam irradiation, the surface of all specimens was subjected to the pre-treatments described in [4]. The samples were irradiated by LEHCPEB on the SOLO setup (Institute of High Current Electronics SB RAS, Tomsk, Russia) at an energy density $E = 10$ J/cm^2, number of pulses $N = 10$, pulse duration $\tau = 50$ µs. The procedure of the formation of the surface alloys on the TiNi substrate is described in detail in [3]. Simultaneous deposition of $Ti_{60}Ta_{40}$ (at.%) coating on the TiNi substrate and pulsed melting of film/substrate system by LEHCPEB was performed on the RITM-SP setup (Institute of High Current Electronics SB RAS, Tomsk) in the following mode: pulse duration $\tau = 2$–2.5 µs, maximum electron energy 17 keV, energy density $E_S = 2.0 \pm 0.2$ J/cm^2, number of pulses $n = 10$. The number of cycles was $N=30$, thus, the predicted thickness of the surface alloy was ~1.5 µm.

TEM experiments were performed on JEM 2100 (JEOL, Japan) electron microscope at the accelerating voltage of 200 kV. Thin foils were prepared by ion thinning. Mechanical tests of the pseudoelastic properties of the TiNi samples (square needles of 1×1×25 mm size) before and after surface treatments were carried out in the loading-unloading cycles with an inverted torsion pendulum. Mechanical torsion tests were carried out at two constant temperatures (T=273 K and T=310 K). When the total deformation reaches 6 %, the sample was unloaded and the inelastic deformation was recovered by the superelasticity. After unloading, the sample was heated above 373 K and the residual strain was recovered due to shape memory effect. Hardness measurements of the specimens was examined by nanoindentation on a CSEM Nano Hardness Tester (USA) with a Vickers pyramid under constant load increased stepwise of 100 mN.

Results and discussion

The surface gradient structure, shown in Fig. 1a, of TiNi alloy after irradiation with LEHCPEB consists of several layers: (i) upper oxide layer of a 30 nm thickness consisting of a mixture of TiO_2 and intermetallic Ti_3Ni_4 nanoparticles; (ii) adjacent sublayer of a 3 µm thickness having a columnar structure on the base of B2 phase; (iii) the underneath sublayer, residing at a depth of ~3 µm and extends to ~5 µm, consists of alternating columnar and globular substructures; (iv) bottom sublayer with an high dislocation density at a depth of more than 5 µm. The columnar B2 grains have 200×500 nm size in the transverse and longitudinal direction. Along the B2 columnar grain boundaries, TiO_2 и Ti_3Ni_4 nanoparticles of size less than 10 nm are segregated.

Fig. 1b shows cross-sectional bright-field TEM image and corresponding microdiffraction patterns of Ti-Ta-Ni surface alloy and an intermediate layer adjacent to TiNi substrate. It can be seen that the outer (near-surface) layer of a ~1.8 µm thickness has uniform amorphous structure. Corresponding diffraction pattern (Fig. 1b, inset) contain a broad diffuse halo. Beneath the amorphous outer layer, the amorphous-crystalline sublayer of a ~500 nm thickness is observed. In corresponding microdiffraction pattern (Fig. 1b, inset), near the diffuse halos, the point Bragg reflections from highly dispersed particles of Ti_2Ni phase are found.

Experimental stress-strain (S-S) curves obtained at T=273 K and T=310 K of the TiNi samples before and after surface treatments are shown in Fig.2. In the tests, shear stress was applied till about 6 % deformation was reached and then the stress was removed. It is seen in the figure that pseudoelastic shape recovery takes place only partially but on heating above 373 K the shape memory is almost complete and the residual deformation vanishes. The S–S curves are typical for pseudoelastic behavior and are divided into well-known stages: (i) linear portion due to elastic deformation; (ii) the plateau stage associated with stress-induced martensitic transformations.

Materials Research Forum LLC
doi: http://dx.doi.org/10.21741/9781644900017-12

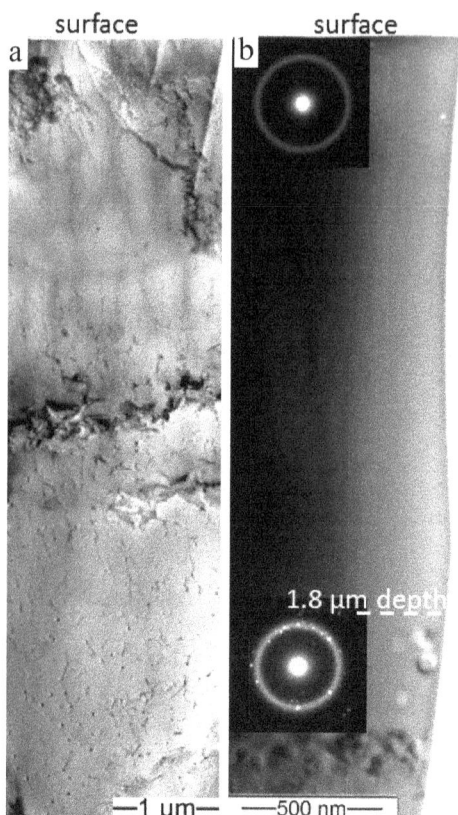

Figure 1. Cross-sectional bright-field TEM images of TiNi alloy treated by LEHCPEB (a) and surface alloy (b). Insets on (b) – microdiffraction patterns obtained from the upper amorphous layer and the amorphous-crystalline sublayer.

At the T=273 K temperature the residual strain recovers by the predominant growth of one variant of B19' martensite and due to rearrangement of the variants of the B19' martensite, accompanied by the twin boundaries movement. It has been found that at both test temperatures, the martensitic shear stress increases for TiNi samples after LEHCPEB treatment, but decreases for the surface alloy. This circumstance in most likely due to the high level of residual stresses induced by multiple electron-beam treatment. For this reason the smalless hysteresis of the surface alloy/TiNi sample related to the ease of interface movement during the tranformation are associated to a larger extent with oriented elastic stress fields. In irradiated with LEHCPEB samples the large thermal hysteresis during the shape recovery are caused by the stress field of the dislocations. It should be mentioned that TiNi samples with modified surface exhibit a higher superelasisity at the low temperature (T = 273 K) than the initial TiNi sample. The elastic strain in the TiNi alloy after LEHCPEB treatment shows significant enhancement and reaches ~1 % whereas the pseudoelastic strain naturally decreases. At high temperature (T=310 K) the alloy is in a single-phase high-temperature superelastic state and superelasticity is realised by the stress-induced martensitic transformation B2→B19'. The martensitic shear stresses increases with increasing temperature that is in accordance with the Clausius-Clapeyron relationship. One-stage stress-induced martensitic transformation is observed without substantial strain hardening, that, as a rule, is realized by the precipitation hardening or introducing plastic strains such as dislocations. In this case, the relative contributions in psuedoelastic deformation, recovered by the shape memory effect and superelasticity, differ depending on the surface treatment. Thus, the samples with surface alloys (to a greater extent) and samples irradiated with LEHCPEB (to a less extent) also exhibit a higher superelasticity strain, compared to the initial TiNi samples. Due to the higher superelasticity strain, the shape memory effect occurs in a less degree. It is important to note that the surface treatments do not lead to the decrease in pseudoelastic charecteristics of the material and additional residual (plastic) deformation does not exceed 0.15 %.

Shape Memory Alloys – SMA 2018   Materials Research Forum LLC
Materials Research Proceedings **9** (2018) 58-62   doi: http://dx.doi.org/10.21741/9781644900017-12

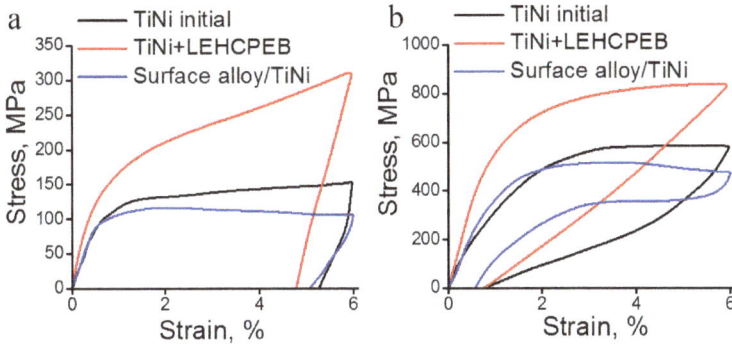

Figure 2. Stress-strain curves obtained at T=273K (a) and T=310 K (b)
of the initial TiNi alloy (black curve), TiNi alloy after LEHCPEB
treatment (red curve) and surface alloy/TiNi (blue curve).

The hardness depth profiles are shown in Fig 3. It has been established that under irradiation the hardness distribution various nonmonotonically as the distance from the surface increases. At the depth from 100 μm and up to 400 μm the continuous hardess maximum is formed and one more maximum is formed at the larger distance from the irradiated surface. The increased microhardness in this sublayers is due to strain hardening under the action of quasistatic stresses formed at the stage of cooling after the completion of resolidification [5]. At average microhardness of the surface layer increased by 20 % (from 2.5 GPa in the initial sample up to 3 GPa). These results adjust with the data described in [3], in which the LEHCPEB treatment is also accompanied by an increase in hardness.

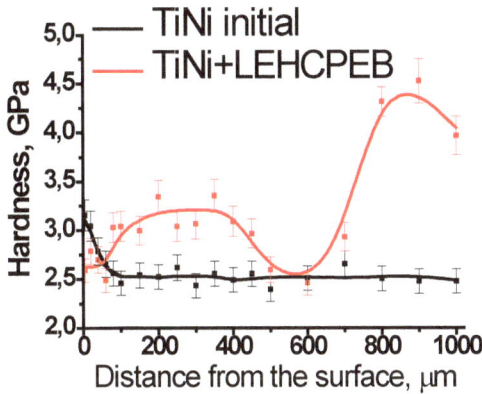

Figure 3. Hardness depth profiles for TiNi alloy before and after LEHCPEB
treatment. The Vickers diamond pyramid load was 100 mN

Summary

The surface treatment of TiNi by LEHCPEB resulted in the formation of a gradient structure in a direction perpendicular to the irradiated surface. It is found that the surface alloy has a layered structure and consists of several sublayers with different types of structure. Studies of the mechanical behavior of TiNi alloys before and after the surface modification have shown that at

Shape Memory Alloys – SMA 2018   Materials Research Forum LLC
Materials Research Proceedings **9** (2018) 58-62   doi: http://dx.doi.org/10.21741/9781644900017-12

different test temperatures corresponding to different structure states of the TiNi alloy, pseudoelastic strain recovering upon unloading and heating takes place.

Acknowledgements

This work was funded by Russian Science Foundation (project No. 18-19-00198 of April 26, 2018).

References

[1] T. Yoneyama, S. Miyazaki (Eds.), Shape memory alloys for biomedical applications, Cambridge, England, 2009.

[2] V. Rotshtein, Y. Ivanov, A. Markov, Surface treatment of materials with low-energy, high-current electron beams, in: Y. Pauleau (Ed.), Materials surface processing by directed energy techniques, Elsevier, Oxford, 2006, pp. 205-240. https://doi.org/10.1016/B978-008044496-3/50007-1

[3] S.N. Meisner, E.V. Yakovlev, V. O. Semin, L. L. Meisner, V.P. Rotshtein, A. A. Neiman, F. D. D'yachenko, Mechanical behavior of Ti-Ta-based surface alloy fabricated on TiNi SMA by pulsed electron-beam melting of film/substrate system, Appl. Surf. Sci. 437 (2018) 217-226. https://doi.org/10.1016/j.apsusc.2017.12.107

[4] A.A. Neiman, L.L. Meisner , A.I. Lotkov, N.N. Koval, V.O. Semin, A.D. Teresov, Cross-sectional TEM analysis of structural phase states in TiNi alloy treated by a low-energy high-current pulsed electron beam, Appl. Surf. Sci. 327 (2015) 321-326. https://doi.org/10.1016/j.apsusc.2014.11.173

[5] A.B. Markov, Mechanisms for modification of carbon steels under the action of a high-energy, high-current electron beam, Russ. Phys. J. 42 (1999) 293-298. https://doi.org/10.1007/BF02508310

Shape Memory Alloys – SMA 2018
Materials Research Proceedings **9** (2018) 63-67

Materials Research Forum LLC
doi: http://dx.doi.org/10.21741/9781644900017-13

Microstructure Formation and Transformation Behavior in Titanium Nickelide with Various Grain Size of B2 Austenite

Elena P. Ryklina[1,a*], Kristina A. Polyakova[1,b],
Natalya Yu. Tabachkova[1,c], Natalia N. Resnina[2,d], Sergey D. Prokoshkin[1,e]

[1]National University of Science and Technology "MISiS", 4, Leninskiy prospect, Moscow 119049, Russia

[2]Saint Petersburg State University, 7/9 Universitetskaya nab., Saint Petersburg, 199034, Russia

[a]ryklina@tmo.misis.ru, [b]vachiyan@yandex.ru, [c]ntabachkova@gmail.com, [d]resnat@mail.ru, [e]prokoshkin@tmo.misis.ru

*corresponding author

Keywords: Shape Memory Alloys, Titanium Nickelide, Martensitic Transformations, Aging, Nanostructures, Morphology of Ni_3Ti_4 Phase

Abstract. The size and morphology of Ti_3Ni_4 precipitates, as well as their transformation behavior after isothermal aging, were studied in a Ti-50.7 at.%Ni shape memory alloy with various B2 austenite grain sizes. The study explains the influence of GS on the precipitated particle size and morphology, transformation kinetics and staging.

Introduction

Among the factors determining the microstructure and martensitic transformations (MT) in Ni-rich titanium nickelide are studied as follows: the Ni content [1], temperature and time of isothermal aging [2], the defects density of structure [3,4]. In [5] the grain size (GS) effect on calorimetric effects was proved and the authors suppose that the GS should also affect the microstructure, but their assumptions are not supported by corresponding research. No other works devoted to a GS effect on microstructure formation and calorimetric effects had been found. To fill a gap in this knowledge a systematic structure investigation should be carried out on Ni-rich alloy with different GSs of B2 austenite subjected to isothermal aging for different durations.

Experimental

The present study was performed using a 0.15 mm thick band of Ti-50.7 at.%Ni alloy obtained as a result of multipass cold rolling (CR) with accumulated strain of $e = 0.6$. Recrystallization annealing (RA) was performed at temperatures of 600 °C (1 h), 700 °C (20 min) and 800 °C (1 h), followed by water quenching in order to obtain the recrystallized structure of B2 austenite with various GSs. Next, the band was cut into 3 mm long specimens and divided into 3 groups for subsequent isothermal aging at 430 °C for 1, 3 and 10 h. For structural analysis, thin foils were prepared by mechanical polishing to a thickness of 0.1 mm and then thinning by ion bombardment using a Gatan PIPS II. Transmission electron microscopy (TEM) observations were carried out using a JEM 2100 operating at 200 kV. The average GS and the parameters of Ti_3Ni_4 particles were determined using the random linear intercept method. The parameters of the particles and the nature of their distribution were estimated in the 0.5 μm wide grain boundary (GB) region and in the grain interior (GI) where the apparent particle length and width (i.e., "diameter" and "thickness"), inter-particle distance and their linear frequency (LF) and volume fraction (VF) were evaluated. A shape factor K was calculated as the diameter-to-thickness (l/h) ratio for the morphological evaluation of particles. Energy dispersive spectroscopy (EDS) was carried out using the EDS stage of a JEOL 2100 using samples with

Published under license by Materials Research Forum LLC.

Shape Memory Alloys – SMA 2018   Materials Research Forum LLC
Materials Research Proceedings 9 (2018) 63-67   doi: http://dx.doi.org/10.21741/9781644900017-13

a homogenized structure obtained as a result of CR with accumulated strain of 0.6 and subsequent RA at 600 °C for 1 h and 800 °C for 1 h. The kinetics of MT was analyzed using differential scanning calorimetry (DSC, Mettler Toledo 822^e calorimeter).

Results and Discussion

After the RA in the temperature range of 600–800 °C the structure of B2 austenite with GSs of 5, 11 and 15 µm is formed. The details of the structure are published in [3].

In the process of subsequent isothermal aging at 430 °C, the Ti_3Ni_4 phase precipitates. The morphology and the degree of large-scale heterogeneity of Ti_3Ni_4 precipitates in the microstructure strongly depends on the GS and aging time. It is poorly expressed after aging for 1 h [6] and pronounced after aging for 3 and 10 h. The microstructural evolution under aging for 10 h in the material with different GSs can be evaluated using TEM images presented in Figure 1. The results of the quantitative phase analysis permit the obtaining of a more definite concept of the GS effect on the character of the precipitation process and confirms the above visual estimation of the microstructure (Table 1).

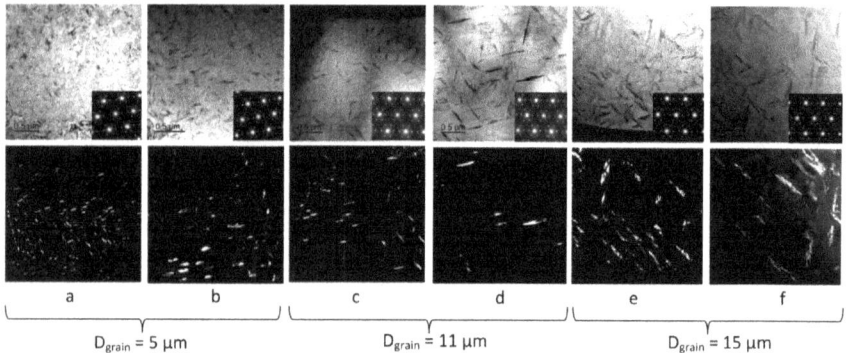

Figure 1. Bright- and dark-field TEM images and SAEDPs (<111>$_{B2}$ zone axis) of Ti-50.7%Ni alloy after aging at 430 °C for 10 h with different initial GSs of B2 austenite in the GB region (a, c, e) and GI (b, d, f).

The particle size is minimal along the GBs and GB region and increases at a distance, as well as *LF*. The form factor *Ff*, which is determined as a ratio of the particle diameter *l* to its thickness *h*, permits tracing of the degree of particle elongation (see Table 1). The fragmented precipitates are observed in all grains, but visually the fragmentation is more expressed in the structure with the largest GS of 15 µm with the largest particles. The presence of fragmented precipitates after aging for 3 and 10 h in the large particles in the large grains (Figs. 2d,f) indicates a loss of their coherency with the B2 matrix [7].

Note that the threefold grain growth from 5 to 15 µm is accompanied by a twofold growth of the particles in the GB region and a threefold growth in the GI. When approaching the GI, the particle shape varies from ellipsoidal to strongly flattened lenticular. The most elongated particles, with a diameter eight times their thickness, are observed in the structure with the GS of 15 µm.

The EDS results show that the distribution of the elements in the 3 µm-grain is quite uniform over the grain cross-section. In the 9-mm grain, the segregation of elements in the GB region becomes pronounced. The Ni content reaches 54 % and that of Ti decreases to 46 %.

Shape Memory Alloys – SMA 2018   Materials Research Forum LLC
Materials Research Proceedings **9** (2018) 63-67   doi: http://dx.doi.org/10.21741/9781644900017-13

Table 1. Size and distribution parameters of Ti_3Ni_4 precipitates in different zones of the grains after annealing at a temperature of 430 °C for 1 and 10 h (A – along GB; B – GB region (0.5 µm); C – GI)

Parameter	Average grain size, [µm]		
	5	11	15
	A/B/C*	A/B/C	A/B/C
Aging 430 °C, 1 h			
l, [nm]	25/36/60	28/36/70	40/53/72
h, [nm]	10/14/15	12/14/18	12/18/18
$F_f = l/h$	2.5/2.5/4.0	2.3/2.0/3.9	3.3/2.7/4.0
LF, [p/µm]	–/8.0/6.0	–/8.0/6.0	–/5.0/4.0
VF,%	–/3.3/2.2	–/2.4/2.3	–/2.4/2.3
Aging 430 °C, 10 h			
l, [nm]	35/100/120	40/100/230	180/200/350
h, [nm]	12/25/35	18/25/35	30/40/45
$F_f = l/h$	2.9/4.0/3.4	2.2/4.0/6.6	6.0/5.0/8.0
LF, [p/µm]	–/6.0/5.0	–/5.0/2.0	–/4.0/1.0
VF, [%]	–/4.05/2.4	–/4.05/2.4	–

This provides an explanation for the difference between the observed microstructures in grains with different GSs. It is known that the defect density is higher in the fine-grained structure [8], therefore, the number of nucleation centers of particles is high at the same concentration of nickel over the grain cross section, no difference in the particle size is observed. In the coarse-grained structure the number of nucleation centers of particles is lower and in the presence of the higher Ni concentration the precipitates grow to the particles grow to larger sizes.

The calorimetric curves corresponding to a structure with different GS and different aging times by the DSC method are presented in Fig. 2.

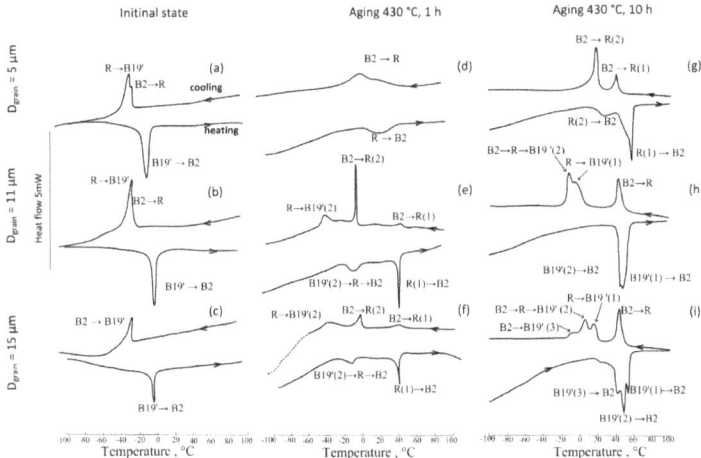

Figure 2. Calorimetric curves of Ti-50.7 at.%Ni alloy with different initial GSs of B2 austenite: hot-rolled (a–c) and after isothermal aging at 430 °C for 1 h (d–f) and 10 h (g–i).

The effect of annealing time on the temperature of the B2→R transformation is not pronounced, its peak remains within a 39–43 °C temperature range. In the fine-grained structure (GS = 5 µm), B19′ martensitic formation is suppressed after aging during 1–10 h; it becomes possible with GS growth above 11 µm and an increase in the degree of heterogeneity of the microstructure. The appearance of a B2→B19′ transformation (without an intermediate R phase) is observed in the structure with the largest GS of 15 μm at the maximum holding time. This is associated with the loss of the coherency between coarse Ti_3Ni_4 precipitates (~350 nm) and the matrix [7].

Conclusions

1. The B2 austenite grain strongly affects the microstructure of Ti_3Ni_4 precipitates formed in the process of isothermal aging at 430 °C for 1–10 h in a Ti-50.7 at% Ni alloy. The distribution of Ti_3Ni_4 precipitates is heterogeneous, the degree of which depends on aging time: it is mild after aging for 1 h and becomes after aging 3 and 10 h. The particle diameter grows from the boundaries towards the grain interior, the shape of the particles becomes more elongated, the linear frequency of their distribution and volume fraction decrease. A three-fold increase in GS from 5 to 15 μm is accompanied by two- and three-fold increases in particle diameter in the grain boundary region and grain interior, respectively.

2. The EDS results estimate the difference in the distribution of Ni and Ti in the grains of different sizes. The elements distribution is quite uniform over the grain cross-section in the fine-grained structure; in the coarse-grained structure, the Ni content reaches 54 % near the grain boundary, which results in the formation of the coarse Ti_3Ni_4 particles after aging for 10 h.

3. The grain size and microstructure strongly affect the staging of martensitic transformation. In the fine-grained structure B19′ martensitic formation is suppressed. The formation of B19′ martensite becomes possible with GS growth above 11 μm and an increase in the degree of heterogeneity of the microstructure. One-stage B2→B19′ transformation is observed in the structure with the largest GS of 15 μm after aging 10 h due to the loss of the coherency between coarse Ti_3Ni_4 precipitates and the matrix.

Acknowledgments

The present work was carried out with financial support from the Ministry of Education and Science of the Russian Federation in the frameworks of State Task No.11.1495.20174.6 and from RFBR according to the research project № 18-08-01193 A.

References

[1] B. Karbakhsh Ravari, S. Farjami, M. Nishida, Effects of Ni concentration and aging conditions on multistage martensitic transformation in aged Ni-rich Ti–Ni alloys, Acta Mater. 69 (2014) 17-29. https://doi.org/10.1016/j.actamat.2014.01.028

[2] J. Khalil-Allafi, G. Eggeler, A. Dlouhy, W.W. Schmahl, Ch. Somsen, On the influence of heterogeneous precipitation on martensitic transformations in a Ni-rich NiTi shape memory alloy, Mater. Sci. Eng. A. 378 (2004) 148-151. https://doi.org/10.1016/j.msea.2003.10.335

[3] K.A. Polyakova-Vachiyan, E.P. Ryklina, S.D. Prokoshkin and S.M. Dubinskii, Dependence of the functional characteristics of thermomechanically processed titanium nickelide on the size of the structural elements of austenite, Phys. Metals Metallogr. 117 (2016) 817-827. https://doi.org/10.1134/S0031918X16080123

[4] A.Yu. Kolobova, E.P. Ryklina, S.D. Prokoshkin, K.E. Inaekyan, V. Brailovskii, Study of the evolution of the structure and kinetics of martensitic transformations in a titanium nickelide upon isothermal annealing after hot helical rolling, Phys. Metals Metallogr. 119 (2018) 134-145. https://doi.org/10.1134/S0031918X17120079

[5] X. Wang, C. Li, B. Verlinden, J. Van Humbeeck, Effect of grain size on aging microstructure as reflected in the transformation behavior of a low-temperature aged Ti–50.8 at.% Ni alloy, Scripta Mater. 69 (2013) 545-548. https://doi.org/10.1016/j.scriptamat.2013.06.023

Shape Memory Alloys – SMA 2018   Materials Research Forum LLC
Materials Research Proceedings **9** (2018) 63-67   doi: http://dx.doi.org/10.21741/9781644900017-13

[6] E.P. Ryklina, K.A. Polyakova, N.Yu. Tabachkova, N.N. Resnina, S.D. Prokoshkin, Effect of B2 austenite grain size and aging time on microstructure and transformation behavior of thermomechanically treated titanium nickelide, J. All. Compd. 764 (2018) 626-638. https://doi.org/10.1016/j.jallcom.2018.06.102

[7] V.I. Zel'dovich, G.A. Sobyanina, V.G. Pushin, V.N Khachin, Phase Transformation in TiNi-Based Alloys. II. Aging during Continuous Cooling, Phys. Metals Metallogr. 77 (1994) 77-81.

[8] S.S. Gorelik, S.V. Dobatkin, L.M. Kaputkina, Recrystallization of metals and alloys, MISiS publ., Moscow, Russia, 2005.

Shape Memory Alloys – SMA 2018
Materials Research Proceedings 9 (2018) 68-73
Materials Research Forum LLC
doi: http://dx.doi.org/10.21741/9781644900017-14

Transmission Electron Microscopy Study of the Atomic Structure of Amorphous Ti-Ta-Ni Surface Alloy

Viktor O. Semin[1,2]

[1]Institute of Strength Physics and Material Science SB RAS, 2/4 Akademicheskii Prospect, Tomsk 634055, Russia

[2]National Research Tomsk State University, 36 Lenin Ave, Tomsk 634050, Russia

lpfreedom14@gmail.com

Keywords: Surface Alloy, Amorphous Structure, Short-Range Order, Radial Distribution Function

Abstract. This paper presents experimental results of studying of the atomic (clustered) structure of an amorphous Ti-Ta-Ni surface alloy using real-space radial distribution functions (RDF) and transmission electron microscopy (TEM). The characterization of the structure by electron diffraction methods has shown that the surface alloy of a 2 μm thickness is completely amorphous without any crystalline phases. Based on the electron diffraction data chemical and topological short-range order and a certain degree of medium-range order in this alloy are found. It is also shown that topological short-range order in the amorphous Ti-Ta-Ni surface alloys is well approximated by the coordination polyhedra based on the intermetallic compounds Ti_2Ni and B2(TiNi): the icosahedron Ti_7Ni_6 and the rhombic dodecahedron Ti_7Ni_8.

Introduction

Thin-film metallic glasses and metallic glasses (MGs) possess extraordinary physical properties (high strength, corrosion resistance, high saturation magnetization), making possible the fabrication of micro and nano devices/instruments for MEMS [1,2] and for biomedical applications [3]. The scientific and technological interest in this group of alloys is caused by the unique nature of the amorphous state associated with the atomic structure of the amorphous materials. However, the relationship between the structure and unusual magnetic, strength and corrosion properties of MGs is still unclear due to the fact that the glassy state is characterized by the absence of atomic long-range order and translational symmetry. Well-known approaches and diffraction methods, which are typical for the description of the crystalline structures, are not applicable for the amorphous structures. Therefore the description of the atomic arrangements in these noncrystalline materials in the three-dimensional space, as well as the nearest neighbor distances or coordination numbers, requires theoretical modeling or employing of the direct experimental methods.

The distinguishing feature of the X-ray, neutron or electron diffraction patterns obtained from MGs is known to be the presence of one or more broad diffuse halos. Averaged distribution of atoms located on the n-th coordination shell can be deduced from the diffraction data by means of radial distribution functions (RDFs) [4]. Although the RDFs provide only statistically averaged data, but using nanometer-beam electron probes and structural models it is possible to reconstruct the real atomic structure of the amorphous material.

In the present work an investigation of atomic (clustered) structure of the amorphous Ti-Ta-Ni surface alloy using real-space RDFs derived from electron nano-beam diffraction patterns was performed.

Published under license by Materials Research Forum LLC.

Shape Memory Alloys – SMA 2018   Materials Research Forum LLC
Materials Research Proceedings **9** (2018) 68-73   doi: http://dx.doi.org/10.21741/9781644900017-14

Experimental work

The formation of the Ti-Ta-Ni based surface alloys was carried out according to the procedure described in [5]. Simultaneous deposition of $Ti_{60}Ta_{40}$ (at. %) thin-film coating on the TiNi substrate and pulsed melting of film/substrate system by low-energy high-current electron beam was performed in the following mode: pulse duration τ=2–2.5 µs, maximum electron energy 17 keV, energy density E_S=2.0 ± 0.2 J/cm^2, number of pulses n=10 and repetition rate of 1 pulse/5 s. The number of cycles was N=10, thus, the predicted thickness of the surface alloy was ~1 µm.

TEM experiments were performed on JEM 2100 and JEM 2100F electron microscopes (JEOL, Japan) at the accelerating voltage of 200 kV. Thin foils were prepared by ion thinning using EM 09100IS device (JEOL, Japan). The resulting reduced electron RDFs G(r) (further in the text will be denoted as RDF) were obtained using routine method [6] by means of nano-beam diffraction (NBD) patterns. This approach is suitable for structural analysis of small material volumes in standard transmission electron microscope with no energy filtering and can be improved, if the high-frequency oscillations of the structure factor $\varphi(q)$ for the scattering vectors $q=4 \cdot \pi \cdot \sin(\theta)/\lambda > 8$ Å^{-1} (θ is the scattering angle and λ is the wavelength of the scattered radiation) are taking into account by applying the damping term $\sim exp(-bq^2)$ [6]:

$$\varphi(q) = \left[\frac{I(q)_{experimental} + \sum_i N_i < f_i^2(q) >}{< f_i(q) >^2} \right] q \cdot \exp(-bq^2), \tag{1}$$

$$G(r) = 4 \int_0^\infty \varphi(q) \sin(qr) dq = 4\pi \left[g(r) - \rho_0 \right], \tag{2}$$

where ρ_0 – the mean density, $f_i(q)$ – the atomic scattering factor, N_i – the amount of scattered atoms. The scattering factor of investigated amorphous Ti-Ta-Ni phase was calculated from the atomic scattering factors of Kirkland [7] with the controlled chemical composition. The local elemental composition of the amorphous regions, from which NBD patterns were collected, was controlled by the energy dispersive spectrometer (EDS) INCA Energy (Oxford Instruments, UK) installed on TEM. NBD patterns were collected at a probe beam diameter of 0.5–2.4 nm using a camera length of 60 cm. The number of NBD patterns coverted to PDFs was at least three. The foils were more than 50 nm thickness therefore the coordanation numbers cannot be calculated due to the multiple and inelastic scattering. Electron diffraction patterns were azimuthally integrated and averaged up to the scattering vectors of $q = 10$ Å^{-1}. The quantitative intensity profile analysis based on NBD patterns was carried out using the PASAD-tools software (designed as a plugin for Digital Micrographe) [8]. Using the above described procedure, the nearest-neighbor distances can be determined to ±0.003 nm.

Results and discussion

According to TEM results, it has been found that the surface alloy fabricated on TiNi substrate is completely amorphous (Fig. 1) and no crystalline phases are observed. Fig. 1d is a typical high-resolution electron microscopy (HREM) image with maze-like disordered structure. The corresponding diffraction patterns (Fig. 1b,d, insets) do not exhibit any spot reflections. Although no long-range order is observed, there are some regions ~1 nm, indicated in Fig. 1d by circles, corresponding to medium-range-order regions. In the upper part of the amorphous surface layer with a thickness up to 500 nm, nanobubbles (Fig. 1b) are detected. The formation of nanobubbles most likely contributes to the densification of the amorphous matrix, i.e. reducing the thermodynamic free volume and hence the energy of the amorphous structure. The thickness of the amorphous surface alloy is 2 µm, which exceeds to nearly twice the total

Shape Memory Alloys – SMA 2018   Materials Research Forum LLC
Materials Research Proceedings **9** (2018) 68-73   doi: http://dx.doi.org/10.21741/9781644900017-14

thickness of the deposited $Ti_{60}Ta_{40}$ film. Therefore, it can be assumed that the amorphous structure was formed as a result of superfast quenching from the melt.

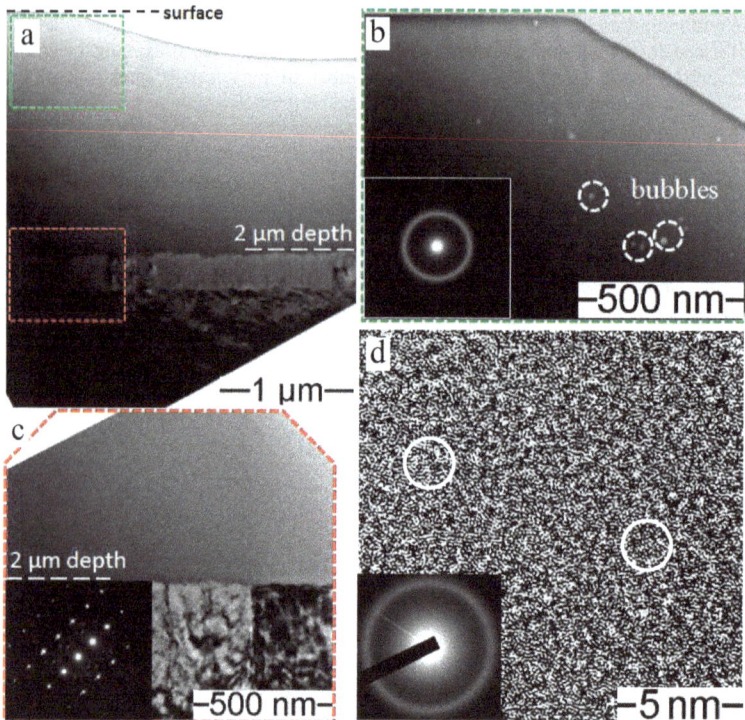

Figure 1. Bright-field TEM images (a, b, c) and high-resolution TEM (d) image of the amorphous surface alloy. Insets on (b-d) – corresponding microdiffraction patterns

The elemental composition by EDS/TEM method (Fig. 2a,d) has shown that during the multiple electron beam treatment of the film/substrate system the upper part (at a depth of 300 nm) of the surface alloy is Ni-rich and has the ternary composition $Ti_{50}Ta_{15}Ni_{35}$ (at.%). The presence of high nickel concentration is due to the high diffusive mobility of nickel atoms that can be easily displaced toward the surface during resolidification from the melt. The more the distance from the surface, the less the tantalum concentration in the surface alloys. The amount of tantalum decreases monotonically and at a depth of 1500 nm the amorphous phase has the composition $Ti_{50}Ni_{43}Ta_7$ (at.%). Thus, the additive fabrication process applied in the present study leads to the formation of an amorphous Ti-Ta-Ni surface alloy with a concentration gradient.

The chemical composition of the amorphous alloy greatly effects on the arrangement of atoms in the first coordination shell, i.e. on the chemical short-range order parameters [4]. Fig. 2b,d shows RDFs of the studied glassy surface alloy, corresponding to the amorphous sublayers at a depth of 300 nm and 1500 nm from the surface, having different chemical composition.

Shape Memory Alloys – SMA 2018   Materials Research Forum LLC
Materials Research Proceedings **9** (2018) 68-73   doi: http://dx.doi.org/10.21741/9781644900017-14

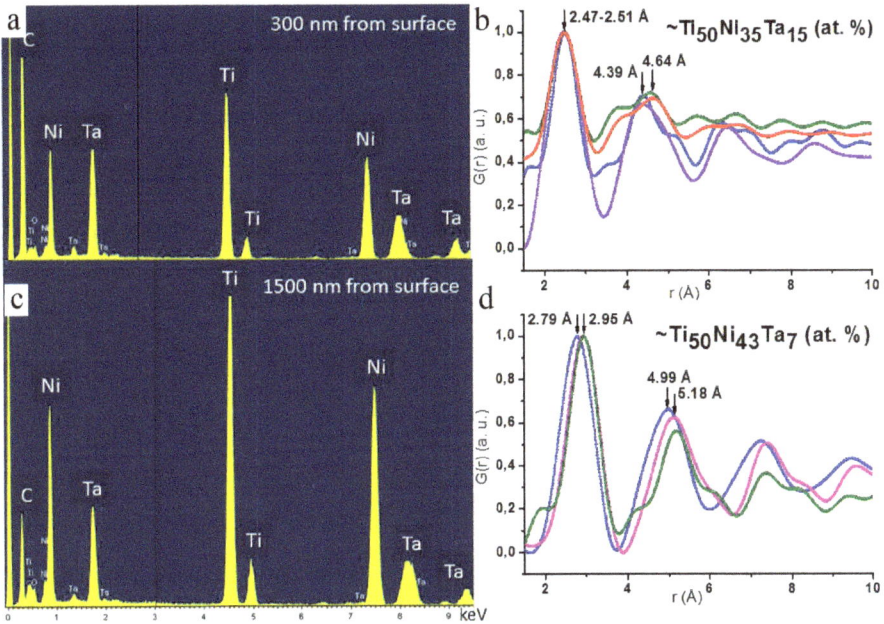

*Figure 2. EDS/TEM spectra, obtained from amorphous sublayers at a depth of 300 nm (a)
and 1500 nm (c) from the surface. RDFs of ternary amorphous Ti-Ta-Ni
surface alloy correspond to amorphous phase at the depth of 300 (b)
and 1500 (d) nm from the surface. Elemental compositions are also given.*

As can be seen from Fig. 2b,d a certain degree of medium-range order in this alloy maintains up to 1 nm distance independently on the chemical composition. The radii of the first (r_1) and second (r_2) coordination shells increase with increasing nickel concentration: from $r_1 = 2.47$–2.51 Å to $r_1 = 2.79$–2.95 Å and from $r_2 = 4.39$–4.64 Å to $r_2 = 4.99$–5.18 Å. Differences in the peak shapes and the positions of the maxima are largely due to the nanometer-sized of analyzed area and the spatial heterogeneities caused by, most likely, local structure variations. Structural information obtained from RDF curves within the first and second coordination shells and (Goldschmidt) atomic and covalent radii are shown on Table 1. Comparison of the experimental data with Goldschmidt bond length (Table 1) allows making some conclusions. Firstly, a strong covalent interaction between the nickel and titanium/tantalum atoms occurs in the first coordination shell in the amorphous phase located at a depth of 300 nm. This is consistent with fact that the Ni-Ti and Ni-Ti interatomic pairs have negative mixing enthalpy. Secondly, the interatomic (Goldschmidt) distance between Ti-(Ti, Ta) is r[Ti-(Ti, Ta)]=2.92 Å and is close related to the position of the main maximum on RDF curves obtained from the amorphous layer at a depth of 1500 nm. It means that the Ni-(Ti,Ta) interatomic pairs are the most predominate atomic pairs in the upper part of the amorphous surface alloy whereas Ti-(Ti,Ta) pairs are prevalent ones at a greater distance from the surface. It shoulde be mentioned that increase in Ni content causes a shift of peaks maxima in the first and second coordination shells toward larger distances. This circumstance contradicts with the fact that the enrichment of the amorphous phase with nickel, having the smallest atomic radius, should lead to the decrease of the interatomic distances in the first coordination shell. To explain this anomaly, it has been assumed that the amorphous matrix contains regions with topological short-range order similar to short-range order of the possible phases of devitrification: Ti_2Ni and B2(TiNi).

Table 1. Structural information obtained from RDFs of the amorphous Ti-Ta-Ni surface alloy and bond lengths in the clusters of Ti_2Ni and B2(TiNi).

Composition, [at.%]	r_1, [Å]	r_2, [Å]
$Ti_{50}Ni_{35}Ta_{15}$	2.47÷2.51	4.39÷4.64
$Ti_{50}Ni_{43}Ta_7$	2.79÷2.95	4.99÷5.18
Atomic pairs	Atomic radius, [Å]	Covalent radius, [Å]
Ni-(Ti,Ta)	2.70	2.47
Ni-Ni	2.48	2.30
Ti-(Ti,Ta)	2.92	2.64
Bonds in Ti_2Ni	First coordination shell, [Å]	Second coordination shell, [Å]
Ti(16c)–Ni	2.45	4.10
–Ti	2.93	3.99
–Ti	–	4.51
–Ni	–	5.20
Bonds in B2(TiNi)	First coordination shell, [Å]	Second coordination shell, [Å]
Ti(1a)–Ni	2.61	5.00
–Ti	3.01	4.26

Fig. 3 shows first coordination polyhedra in the Ti_2Ni structure type (space group Fd3m) at the 16(c) crystallographic position and in the CsCl structure type (space group Pm3m) at 1(a) crystallographic position. The bond lengths in the first coordination clusters are also presented in Table 1. The interatomic distances were obtained using VESTA software allowing to reconstruct 3D structures to determine interatomic distances in the hard sphere model. The positions of atoms (Wyckoff positions) and the lattice parameters can be found elsewhere [9,10].

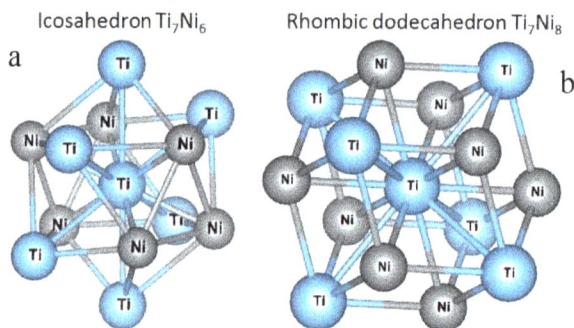

Figure 3. The structure of the first coordination clusters of the Ti_2Ni (a) and B2(TiNi) (b) at the 16(c) and 1(a) crystallographic positions.

Comparison of interatomic distances (Table 1) in clusters with experimental ones (Fig. 2b,d) in the amorphous Ti-Ta-Ni surface alloy leads the following conclusion. The topological short-range order, observed in the amorphous layer at a depth of 300 nm, occurs in the clusters of the icosahedral type Ti_7Ni_6 (Fig. 3a) of Ti_2Ni compound. In turn, the rhombic dodecahedron Ti_7Ni_8 in the B2(TiNi) first-coordination cluster (Fig. 3b) can be used as an approximant for the amorphous phase at a depth of 1500 nm. Various Ti-centered clusters in the first coordination shell reconstruct medium-range orientational ordering if adjacent clusters share solvent atoms in

Shape Memory Alloys – SMA 2018   Materials Research Forum LLC
Materials Research Proceedings **9** (2018) 68-73   doi: http://dx.doi.org/10.21741/9781644900017-14

common faces, edges or vertices. Thus, the proposed version of the cluster structure of the amorphous Ti-Ta-Ni surface alloy consisting of the Ti_7Ni_6 icosahedron and the rhombic dodecahedron Ti_7Ni_8, can describe the atomic medium-range order in the amorphous phase.

Acknowledgements

The author sincerely thanks V.P. Rotshtein, A.B. Markov, G.E. Ozur, E.V. Yakovlev for carrying out the surface treatments and prof. L.L. Meisner for appreciate fruitful discussions. This work was funded by Russian Science Foundation (project No. 18-19-00198 of April 26, 2018).

Summary

The construction of the RDFs $G(r)$ using electron diffraction data is a promising method for studying of the atomic short-range order in the amorphous phase characterized by an inhomogeneous chemical composition. It was shown experimentally that the topological short-range order in the amorphous Ti-Ta-Ni surface alloy is well described by the structures of coordination polyhedra based on the intermetallic compounds Ti_2Ni and B2(TiNi).

References

[1] T. A. Phan, M. Hara, H. Oguchi, H. Kuwano Current sensors using Fe–B–Nd–Nb magnetic metallic glass micro-cantilevers, Microel. Eng. 135 (2015) 28-31. https://doi.org/10.1016/j.mee.2015.02.043

[2] Y. Hayashi, H. Yamazaki, D. Ono, K. Masunishi, T. Ikehashi, Investigation of PdCuSi metallic glass film for hysteresis-free and fast response capacitive MEMS hydrogen sensors, Int. J. Hydrog. Energy 43 (2018) 9438-9445. https://doi.org/10.1016/j.mee.2015.02.043

[3] J.P. Chu, T.-Y. Liu, C.-L. Li, C.-H. Wang, J.S.C. Jang, M.-J. Chen, S.-H. Chang, W.-C. Huang, Fabrication and characterizations of thin film metallic glasses: Antibacterial property and durability study for medical application, Thin Solid Films. 561 (2014) 102-107. https://doi.org/10.1016/j.tsf.2013.08.111

[4] F.E. Luborsky (Ed.), Amorphous Metallic Alloys, first ed., Butterworths, London, 1983. https://doi.org/10.1016/B978-0-408-11030-3.50006-6

[5] L.L. Meisner, A.B. Markov, V.P. Rotshtein , G.E. Ozur, S.N. Meisner, E.V. Yakovlev, V.O. Semin, Yu.P. Mironov, T.M. Poletika, S.L. Girsova, D.A. Shepel, Microstructural characterization of Ti-Ta-based surface alloy fabricated on TiNi SMA by additive pulsed electron-beam melting of film/substrate system, J. Alloys Comp. 730 (2018) 376-385. https://doi.org/10.1016/j.jallcom.2017.09.238

[6] D.J.H. Cockayne, The study of nanovolumes of amorphous materials using electron scattering, Annu. Rev. Mater. Res. 37 (2007) 159-187. https://doi.org/10.1146/annurev.matsci.35.082803.103337

[7] E.J. Kirkland, Advanced computing in electron microscopy, Plenum Press, 1998. https://doi.org/10.1007/978-1-4757-4406-4

[8] C. Gammer, C.Mangler, C.Rentenberger, H.P.Karnthaler, Quantitative local profile analysis of nanomaterials by electron diffraction, Scripta Mater. 63 (2010) 312-315. https://doi.org/10.1016/j.scriptamat.2010.04.019

[9] G.A. Yurko, J.W. Barton, J.G. Parr,The crystal structure of Ti_2Ni, Acta Cryst. 12 (1959) 909-911. https://doi.org/10.1107/S0365110X59002559

[10] K. Otsuka, X. Ren, Physical metallurgy of Ti-Ni-based shape memory alloys, Progr. Mater. Sci. 50 (2005) 511-678. https://doi.org/10.1016/j.pmatsci.2004.10.001

Shape Memory Alloys – SMA 2018
Materials Research Proceedings 9 (2018) 74-79

Materials Research Forum LLC
doi: http://dx.doi.org/10.21741/9781644900017-15

Surface Modification of Ti-Nb-Zr Foams by Poly(3-Hydroxybutyrate)

Vadim A. Sheremetyev[1,a], Anton P. Bonartsev[2,b], Sergey M. Dubinskiy[1,c],
Yulia S. Zhukova[1,d], Garina A. Bonartseva[3,e], Tatiana K. Makhina[3,f],
Elizaveta A. Akoulina[3,g], Elina V. Ivanova[2,h], Maria S. Kotlyarova[2,i],
Sergey D. Prokoshkin[1,j*], Vladimir Brailovski[4,k], Konstantin V. Shaitan[2,l]

[1]National University of Science and Technology "MISiS", 4 Leninskiy Prospect, Moscow
119049, Russia

[2]Faculty of Biology, Lomonosov Moscow State University, 1-12 Leninskie Gory, Moscow
119234, Russia

[3]A.N. Bach Institute of Biochemistry, Research Center of Biotechnology RAS, 33-2 Leninskiy
prosp., Moscow, 119071, Russia

[4]Ecole de technologie superieure, 1100, Notre-Dame Str. West, Montreal (Quebec), H3C 1K3,
Canada

[a]sheremetyev@misis.ru, [b]ant_bonar@mail.ru, [c]dubinskiy@tmo.misis.ru, [d]zhukova@misis.ru,
[e]bonar@inbi.ras.ru, [f]tat.makhina@gmail.com, [g]akoulinaliza@gmail.com, [h]eliza92@yandex.ru,
[i]kotlyarova.ms@gmail.com, [j]prokoshkin@tmo.misis.ru, [k]vbrailovski@mec.etsmtl.ca,
[l]shaytan49@yandex.ru

*corresponding author

Keywords: Biomaterials, Shape Memory Materials, Porous Materials, Polymers, Surfaces

Abstract In this study, Ti-Nb-Zr superelastic foams were produced, characterized from the standpoint of their morphology and mechanical properties. To improve biocompatibility of these foams, they were subjected to surface modification by Poly(3-Hydroxybutyrate). The two-stage immersion of the Ti-Nb-Zr foams in the PHB-containing solution allows forming on their surface continuous polymer layers with incorporation of 6.4 % (w/w) of PHB.

Introduction

Over the past decades, intensive research has been carried out in the field of binary and multicomponent nickel-free superelastic titanium alloys, in particular of Ti-Nb, Ti-Nb-Zr alloy systems, as perspective materials for bone replacement [1-4]. These shape memory alloys (SMAs) demonstrate a unique combination of low Young's modulus (as low as 60–80 GPa), superelastic behavior, which is close to the behavior of bone, and contain only non-toxic elements, such as Ti, Nb, and Zr, in their chemical composition [5].

Even though the Young's modulus of bulk Ti-Nb-based SMAs is low as compared to other metallic biomaterials, it is still significantly higher than that of human bone [1,3]. This mechanical mismatch between the implant and bone leads to the "stress shielding" phenomenon, which is known to be the cause of bone resorption and implant loosening [1]. The idea to use metallic foams, the stiffness of which would be much lower than that of a bulk material is considered to be a promising solution to the problem of stiffness mismatch. In addition, porous structures allow bone ingrowth and, consequently, provide more efficient implant/bone fixation [6].

Recently, the space holder method has been applied to fabricate Ti-based foams for biomedical application [6-9]. This method allows manufacturing open-porosity foams with

Published under license by Materials Research Forum LLC.

different porosity (up to 80 %) and the Young's modulus matching that of bones (1–30 GPa) [6]. Complex architecture of such foams makes extremely challenging the application of conventional surface modification techniques to create on their surface a "human body friendly" environment favorable for bone ingrowth. An exploratory study of this issue constitutes the main objective this work.

Poly(3-hydroxybutyrate), the basic polymer homologue of the polyhydroxyalkanoates' family (PHA), is the most common microbial polyester that has been used as a perspective biodegradable alternative to synthetic thermoplastics. Since PHB manifests simultaneously the biodegradable and biocompatible properties, it has received much attention as the base component for perspective medical devices and drug dosage formulations. PHB can also be used for the surface modification of metallic medical devices by coating and filling – to improve their biocompatibility and provide close integration of the device with a living tissue, e.g. bone tissue [10].

Experimental

Preparation of the Ti-Nb-Zr-foams. A 50 mm-diameter, 600 mm-long Ti-20.8at%Nb-5.5at%Zr (TNZ) ingot was manufactured by *Flowserve Corp.* (USA) and atomized by *TLS Technik Spezialpulver* (Germany). TNZ foams were fabricated using a powder metallurgy based technique called the "space holder process" described in detail in [9]. As a result, 15 mm-diameter, 15 mm-high cylindrical samples of ~50 % ($P \approx 0.5$) porosity foams were obtained. For the surface modification study, these samples were EDM-cut into 1 mm-thick disks, with their centre-line axis either parallel or perpendicular to the compaction direction.

Production of PHB. PHB was produced by bacterial biosynthesis [11]. A PHA producer *Azotobacter chroococcum* strain 7B, a non-symbiotic nitrogen-fixing bacterium able to overproduce PHB (to 80 % of cell dry weight) was used. The strain was isolated from the wheat rhizosphere (sod-podzolic soil) and maintained on Ashby's medium. For PHB synthesis in cells, the culture was grown in shaker flasks (containing 100 ml of the medium) at 30 °C in Burk's medium, containing sucrose as the primary carbon source. Strain growth and polymer accumulation was controlled by nephelometry and light microscopy. The polymer isolation and purification from *A. chroococcum* comprised the following stages: (1) polymer extraction with chloroform in a shaker for 12 hours at 37 °C; (2) separation of polymer solutions from cell debris by filtration; (3) polymer precipitation from chloroform solution with isopropanol; (4) subsequent repeated cycles of dissolution in chloroform and precipitation with isopropanol for 4–5 times to remove any additives and contaminants, and (5) drying at 60°C. Details of PHB and its copolymers biotechnological production have been published in [11].

Modification of metallic foams with polymer. The technique of multiple impregnation of a porous metallic sample in a 1 % polymer solution of PHB in chloroform for several days was used. Soaking was carried out in one stage and in two stages, when the polymer impregnated sample was dried and then again placed in a polymer solution. After soaking, the samples were incubated in distilled water for 2 hours to determine the effectiveness of polymer deposition on the metallic substrate. The entry of PHB into the metallic foams was measured by weighing.

Characterization. The foams' porosity (P), pore size and distribution were characterized using two techniques: a combination of metallography (*NMM-800TRF* optical microscope) and image processing (*ImageJ* software) technique [9], and the independent Archimedes porosity measurement technique [12]. Six images at different locations of each specimen were processed by *Image J* software one by one. The pore sizes range from micro-pores (<10 μm) to macro-pores (from 100 to 1000 μm). The effect of the pore size on the sample architecture is analysed in terms of the ratio "volume of an individual pore to the total volume of pores", which is also

Shape Memory Alloys – SMA 2018   Materials Research Forum LLC
Materials Research Proceedings **9** (2018) 74-79   doi: http://dx.doi.org/10.21741/9781644900017-15

termed as "point impact". The cumulative impact is calculated by summarizing the volume of pores in different diameter ranges, divided by the total pore volume in the sample.

The cyclic compression tests with incrementally increased engineering strain up to either $\varepsilon=50$ % or specimen failure (ε at each cycle corresponds to 0.02 of the initial sample height) were performed using an *MTS' Alliance RF/200* ($\xi=0.002$ s^{-1}).

Polymer coating of metallic foams was studied by scanning electron microscopy (SEM). For the SEM investigation, the samples (uncoated Ti-Nb-Zr-foams, metallic foams coated by one stage and two stages technique in polymer solution soaking) were mounted on aluminium stumps, coated with gold in a sputtering device for 15 min at 15 mA (*IB-3,Giko*) and examined under a scanning electron microscope (*JSM-6380LA, JEOL*).

Results and discussion

Typical image of pore structure of Ti-Nb-Zr sample ($P=0.49$) is presented in Fig. 1a. It can be seen that even though the number of pores smaller than 90 µm is high (Fig. 1b), their collective impact is low. High collective impact of the pores bigger than 90 µm was observed (Fig. 1b).

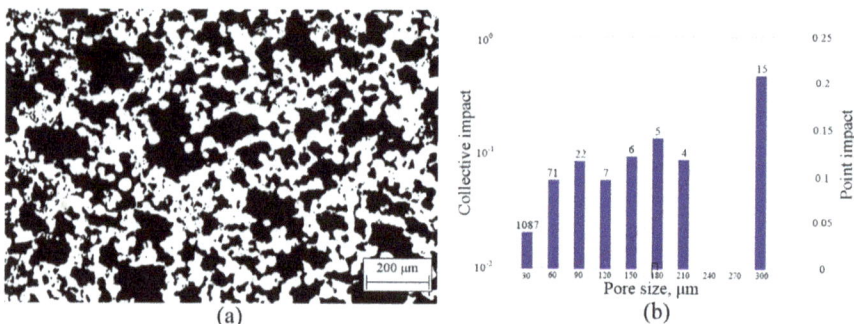

Figure 1. Typical binarized image of pore structure (a) and pore size distribution diagram (b) of Ti-Nb-Zr sample (P=0.49).

Mechanical characterization of fabricated foams was carried out by cyclic compressing tests. A typical stress-strain diagram is presented in Fig. 2. The following parameters were extracted from this diagram: the yield stress (σ_y^*), the engineering stress at $\varepsilon=40$ % (σ_{20}^*), and the Young's modulus at $\varepsilon=20$ % (E^*). The yield stress σ_y^* corresponds to the intersection of the tangent lines to the elastic and the plateau regions [13]. The apparent Young's modulus is determined for the 10th testing cycle ($\varepsilon=20$ %) from the tangent to the point of maximum stress (σ_{20}^*) on the unloading portion of the stress–strain diagram (Fig. 2). The Young's modulus is assessed at an intermediate level of strain to avoid both the influence of the specimen geometry (small strains) and the foam compaction (large strains) on the measured values.

The manufactured TNZ foams ($P=0.49$) demonstrate an interesting combination of low Young's modulus ($E^*=5.7$ GPa), which is close to that of bone, with high yield stress ($\sigma_y^*=214$ MPa), which is more than twice as high as that of bone (Fig. 2) [3].

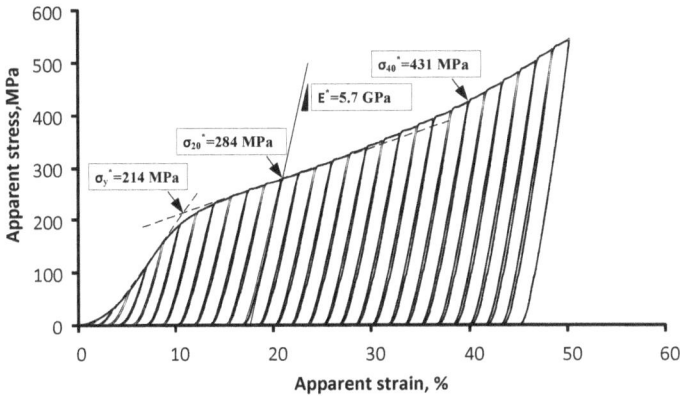

Figure 2. Typical stress-strain diagram of Ti-Nb-Zr foam (P=0.49).

PHB with a molecular weight of 3.55×10^5 Da was produced by bacterial biosynthesis. The coating of Ti-Nb-Zr foams with PHB by one stage and two stages technique of metal foams soaking in polymer solution resulted in the formation of the PHB coating on the surface and inside of pores of the Ti-Nb-Zr foam samples (Fig. 3).

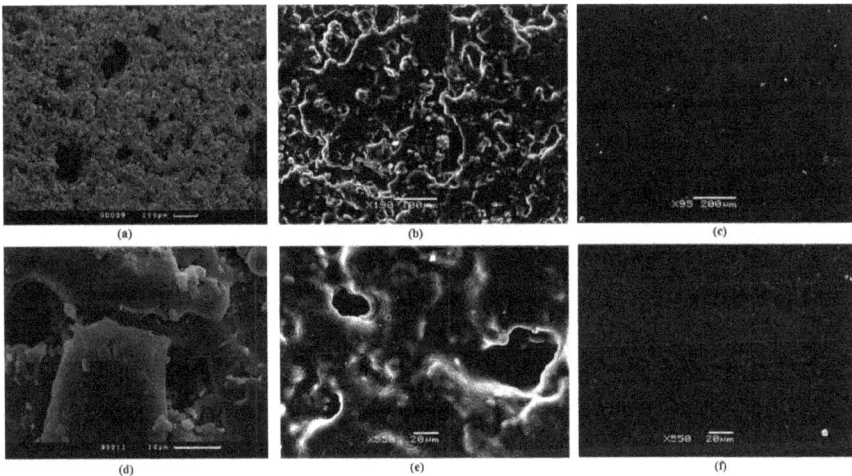

Figure 3. Ti-Nb-Zr foams (a, d) coated with PHB by one stage (b, e) and two stages (c, f) technique of multiple impregnation in polymer solution of Ti-Nb-Zr foam samples; SEM, ×95–190 (a,b,c), ×550–2000 (d, e, f).

Use of the one stage technique of soaking the Ti-Nb-Zr foams in polymer solution led to the formation of a thin and discontinuous polymer layer with a 4.6 % (w/w) content of PHB. The use of the two stage technique allows forming a continuous polymer layer on the surface of Ti-Nb-Zr foam with incorporation of 6.4 % (w/w) of PHB. The incubation of the Ti-Nb-Zr foams coated with PHB in water did not cause detachment of the polymer from the metal substrate, the weight of coated devices did not change.

Conclusion

Thus, this study showed some preliminary results of the successful surface modification of Ti-Nb-Zr foams via the two-stage immersion of these foams in a PHB-containing solution. These results constitute a starting point for the deeper investigation of the applicability of novel surface modification techniques to superelastic Ni-free foams for bone replacement.

Acknowledgments

The present work was carried out with the financial support of the Natural Science and Engineering Research Council of Canada and the Ministry of Education and Science of the Russian Federation (Project ID RFMEFI57517X0158) in part of Ti-Nb-Zr foams production. The equipment of User Facilities Center of Moscow State University (incl. in framework of Development Program of MSU to 2020) and Research Center of Biotechnology RAS was used in the work.

References

[1] M. Geetha, A.K. Singh, R. Asokamani, A.K. Gogia, Ti based biomaterials, the ultimate choice for orthopaedic implants – A review. Prog. in Mater. Sci. 54(3) (2009) 397-425. https://doi.org/10.1016/j.pmatsci.2008.06.004

[2] S. Miyazaki, H.Y. Kim, H. Hosoda, Development and characterization of Ni-free Ti-base shape memory and superelastic alloys, Mater. Sci. and Eng.: A. 438 (2006) 18-24. https://doi.org/10.1016/j.msea.2006.02.054

[3] V. Brailovski, S. Prokoshkin, M. Gauthier, K. Inaekyan, S. Dubinskiy, M. Petrzhik, M. Filonov. Bulk and porous metastable beta Ti–Nb–Zr(Ta) alloys for biomedical applications, Mater. Sci. and Eng.: C. 31 (2011) 643-657. https://doi.org/10.1016/j.msec.2010.12.008

[4] V. Sheremetyev, V. Brailovski, S.Prokoshkin, K. Inaekyan, S. Dubinskiy, Functional fatigue behavior of superelastic beta Ti-22Nb-6Zr(at%) alloy for load-bearing biomedical applications, Mater. Sci. and Eng.: C. 58 (2016) 935-944. https://doi.org/10.1016/j.msec.2015.09.060

[5] M. Niinomi, Recent titanium R&D for biomedical applications in Japan, JOM. 51 (1999) 32-34. https://doi.org/10.1007/s11837-999-0091-x

[6] G. Lewis, Properties of open-cell porous metals and alloys for orthopaedic applications, J. Mater. Sci.: Mater. in Med. 24 (2013) 2293-2325. https://doi.org/10.1007/s10856-013-4998-y

[7] X. Wang, Y. Li, J. Xiong, P. D. Hodgson, C. Wen. Porous TiNbZr alloy scaffolds for biomedical application, Acta Biomater. 5(9) (2009) 3616-3624. https://doi.org/10.1016/j.actbio.2009.06.002

[8] W. Niu, C. Bai, G. Qiu, Q. Wang. Processing and properties of porous titanium using space holder technique, Mater. Sci. and Eng.: A 506 (2009) 148-151. https://doi.org/10.1016/j.msea.2008.11.022

[9] J. Rivard, V. Brailovski, S. Dubinskiy, S. Prokoshkin, Fabrication, morphology and mechanical properties of Ti and metastable Ti-based alloy foams for biomedical applications, Mater. Sci. and Eng.: C. 45 (2014) 421-433. https://doi.org/10.1016/j.msec.2014.09.033

[10] A.P. Bonartsev, S.G. Yakovlev, E.V. Filatova, G.M. Soboleva, T.K. Makhina, G.A. Bonartseva, K.V. Shaitan, V.O. Popov, M.P. Kirpichnikov, Sustained release of the antitumor

drug paclitaxel from poly(3-hydroxybutyrate)-based microspheres. Bioch. (Moscow) Suppl. Ser. B: Biomed. Chem. 6 (2012) 42-47.

[11] A.P. Bonartsev, I.I. Zharkova, S.G. Yakovlev, V.L. Myshkina, T.K. Mahina, V.V. Voinova, A.L. Zernov, V.A. Zhuikov, E.A. Akoulina, E.V. Ivanova, E.S. Kuznetsova, K.V. Shaitan, G.A. Bonartseva, Biosynthesis of poly(3-hydroxybutyrate) copolymers by Azotobacter chroococcum 7B: A precursor feeding strategy, Prep. Biochem. and Biotech. 47 (2017) 173-184. https://doi.org/10.1080/10826068.2016.1188317

[12] Standard Test Methods for Apparent Porosity, Water Absorption, Apparent Specific Gravity, and Bulk Density of Burned Refractory Brick and Shapes by Boiling Water, ASTM International, West Conshohocken (PA) (2010), p. 3

[13] L. Peroni, M. Avalle, M. Peroni, The mechanical behaviour of aluminium foam structures in different loading conditions, Inter. J. Imp. Eng. 35 (2008) 644-658. https://doi.org/10.1016/j.ijimpeng.2007.02.007

Shape Memory Alloys – SMA 2018

Materials Research Forum LLC

Materials Research Proceedings **9** (2018)

doi: http://dx.doi.org/10.21741/9781644900017

The Theory of Martensitic Transformations and Shape Memory Effect: Modeling and Calculations

Shape Memory Alloys – SMA 2018
Materials Research Proceedings 9 (2018) 83-91

Materials Research Forum LLC
doi: http://dx.doi.org/10.21741/9781644900017-16

Elastic Properties of Heusler Alloys Ni(Co)-Mn(Cr, C)-In and Ni(Co)-Mn(Cr, C)-Sn

Danil R. Baigutlin[1,a*], Mikhail A. Zagrebin[1,2,b], Vladimir V. Sokolovskiy[1,3,c],
Vasiliy D. Buchelnikov[1,3,d]

[1]Chelyabinsk State University, 129 Brat'ev Kashirinykh Str., Chelyabinsk 454001, Russia

[2]National Research South Ural State University, 76 Lenin Prospect,
Chelyabinsk 454080, Russia

[3]National University of Science and Technology "MISiS", 4 Leninskiy Prospect,
Moscow 119049, Russia

[a]d0nik1996@mail.ru, [b]miczag@mail.ru, [c]vsokolovsky84@mail.ru, [d]buche@csu.ru

*corresponding author

Keywords: Elastic Constants, Debye Temperature, *Ab Initio* Calculation, Poisson's Ratio, Young's Modulus

Abstract. In this paper, a first-principles study of the structural and elastic Heusler alloys of the form Ni(Co)-Mn(Cr, C)-In and Ni(Co)-Mn(Cr, C)-Sn was made. For the investigated alloys tensors of elastic constants using the density functional theory realized in the VASP package and the finite strain method were calculated. Also, the shear moduli and bulk moduli for polycrystals were calculated using the Hill averaging and Young's moduli. The Poisson's coefficients and Debye temperatures are determined. The shear moduli and bulk moduli for polycrystals using Hill averaging were also calculated and the values of the characteristics such as Young's modulus, the Poisson's ratio and the Debye temperature were calculated.

Introduction

Since its discovery, Heusler's alloys have attracted considerable scientific interest due to the variety of important properties that are manifested, for example, shape memory effect, superelasticity, giant magnetocaloric effect, giant magnetoresistance, giant magnetic deformation, etc. [1,2]. Each of these properties represents both an independent scientific interest and a perspective for practical application. All these properties are associated with the martensitic transformation. A number of experimental studies show that the martensitic transition is accompanied by a pronounced softening of the shear modulus of the initial phase [3-5]. There are also theoretical works on the study of elastic moduli for nonstoichiometric Ni_2MnGa in austenite and martensitic phases [6,7]. To date, the family of Ni_2MnGa alloys, indeed, is one of the most studied families of intermetallic compounds. However, these systems have some drawbacks, such as the high cost of gallium, the low temperature of the martensitic transformation and the Curie temperature below 100 °C [2]. Therefore, it is promising to study other families of Heusler alloys.

This paper presents first-principles studies of the structural and elastic properties of Heusler alloys of the type Ni(Co)-Mn(Cr, C)-In and Ni(Co)-Mn(Cr, C)-Sn.

Computational details

To calculate the components of the elastic constants tensor, it is necessary to do the analysis of the calculated total energy of a crystal as a function of applied strain. To obtain this dependence, consider the energy decomposition of a deformed crystal from the small value of the deformation parameter δ (in this paper not more than 3 %). This is true in the scope of Hooke's law. We have

Published under license by Materials Research Forum LLC.

$$E(V,\varepsilon) = E(V_0,0) + \frac{V_0}{2}\sum_{i,j=1}^{6} c_{ij}\varepsilon_i\varepsilon_j + 0(\varepsilon^3). \tag{1}$$

Using this equation and the corresponding deformation tensors indicated in Table 1, we can obtain the values of the individual elastic constants (C_{ij}).

The presented energy calculations are performed within the framework of the density functional theory (DFT) and the generalized gradient approximation (GGA) using the VASP software package [8].

Table 1. The components of the strain tensor and the corresponding total energy

Name	Components of the strain tensor	$\dfrac{\Delta E(\delta)}{V_0}$
C_1	$\varepsilon_1 = \varepsilon_2 = \varepsilon_3 = \delta$	$\dfrac{3}{2}\left(C_{11} + 2C_{12}\right)\delta^2$
C_2	$\varepsilon_1 = \delta, \varepsilon_2 = -\delta, \varepsilon_3 = \dfrac{\delta^2}{(1-\delta^2)}$	$\left(C_{11} - C_{12}\right)\delta^2 + O(\delta^4)$
C_3	$\varepsilon_3 = \dfrac{\delta^2}{(1-\delta^2)}, \varepsilon_6 = \delta$	$2C_{44}\delta^2 + O(\delta^4)$
T_1	$\varepsilon_1 = \dfrac{\delta^2}{(1-\delta^2)}, \varepsilon_4 = \delta$	$2C_{44}\delta^2 + O(\delta^4)$
T_2	$\varepsilon_3 = \dfrac{\delta^2}{(1-\delta^2)}, \varepsilon_6 = \delta$	$2C_{66}\delta^2 + O(\delta^4)$
T_3	$\varepsilon_1 = \delta, \varepsilon_2 = -\delta, \varepsilon_3 = \dfrac{\delta^2}{(1-\delta^2)}$	$\left(C_{11} - C_{12}\right)\delta^2 + O(\delta^4)$
T_4	$\varepsilon_1 = \delta, \varepsilon_2 = \dfrac{\delta^2}{(1-\delta^2)}, \varepsilon_3 = -\delta$	$\dfrac{1}{2}\left(C_{11} - 2C_{13} + C_{33}\right)\delta^2 + O(\delta^4)$
T_5	$\varepsilon_1 = \varepsilon_2 = \varepsilon_3 = \delta$	$\dfrac{1}{2}\left(C_{11} + C_{12} + 2C_{13} + C_{33}\right)\delta^2 + O(\delta^4)$
T_6	$\varepsilon_3 = \delta$	$\dfrac{1}{2}C_{33}\delta^2 \,(1/2)(C_{33})\delta^2$

To model the chemical disorder, the supercell method was used. In this paper we used a supercell containing 32 atoms. When optimizing the geometry of the supercell, the first Brillouin zone was divided into a $12\times12\times12$ grid according to the Morhorst-Pack scheme [9]. The energy of cut-off of plane waves in calculations was taken equal to 400 eV. The calculations used the PAW PBE potentials [10] with the following electronic configurations: Ni($3d^83p^64s^2$), Co($3d^84s^1$), Mn($3p^63d^54s^2$), C($2s^22p^2$), Cr($3p^63d^54s^1$), In($4d^{10}5p^15s^2$), Sn($4d^{10}5p^25s^2$). The accuracy of calculating the total energy was 10^{-6} eV.

Results and discussion

The elastic constants calculated in this work were performed for a minimized crystal structure. As is known, for cubic crystals there are three independent elements of the tensor of elastic constants, and for tetragonal crystals there are six. These constants were determined by applying deformation to an undistorted cell and solving the corresponding system of equations. The

Shape Memory Alloys – SMA 2018   Materials Research Forum LLC
Materials Research Proceedings **9** (2018) 83-91   doi: http://dx.doi.org/10.21741/9781644900017-16

deformation parameter (δ), in these calculations, took values from –0.03 to 0.03 in increments of 0.01.

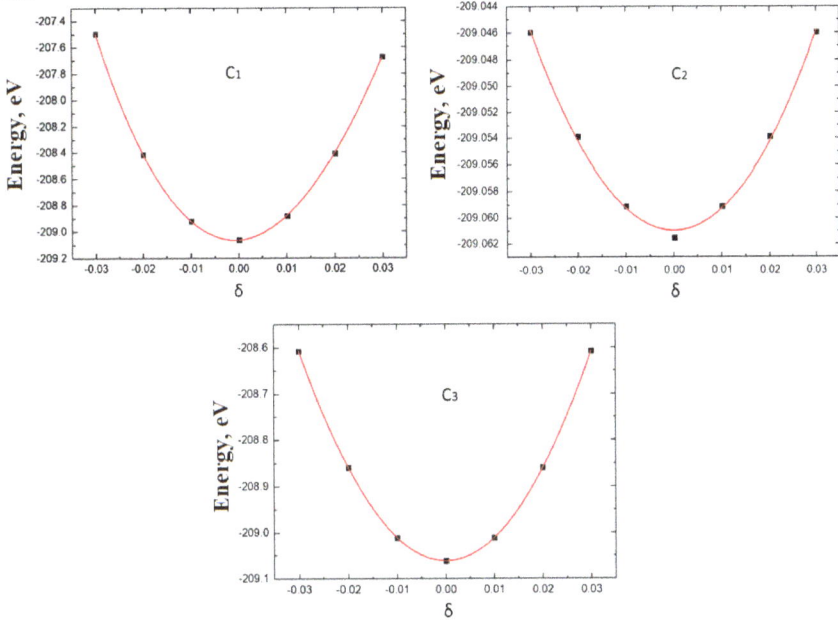

Figure 1. The curve of the total energy of the alloy as a function of the deformation for the strain tensors C_1, C_2, C_3 for the $Ni_{16}Mn_{12}In_4$ alloy. The points are the calculated values of the energy, the solid red line is the approximation of the polynomial of the second degree.

For example, Fig. 1 shows the dependence of the total energy of the alloy on the strain parameter $-\Delta E/V_0(\delta)$ for the $Ni_{16}Mn_{12}In_4$ alloy. These dependencies, for each investigated alloy, were approximated by the least-square method by polynomials of the second degree. The individual elastic constants were calculated from the coefficients of the second powers of the polynomials by solving the corresponding systems of equations presented in Table 1.

The calculated elastic constants for the cubic phase are shown in Table 2. Also, the bulk modulus and the shear modulus were calculated, using the following known relationships:

$$B = \frac{c_{11} + 2c_{12}}{3}, \tag{2}$$

$$c' = \frac{c_{11} - c_{12}}{2}. \tag{3}$$

Shape Memory Alloys – SMA 2018   Materials Research Forum LLC
Materials Research Proceedings **9** (2018) 83-91   doi: http://dx.doi.org/10.21741/9781644900017-16

*Table 2. The equilibrium lattice constant [Å], the elastic constants (C_{11}, C_{12}, C_{44} [GPa]),
the bulk moduli (B [GPa]), and the shear modulus (C' [GPa]) of the investigated alloys
in the cubic phase (the results with asterisk (*) are obtained by the EMTO method in [11])*

Name	a	C_{11}	C_{12}	C_{44}	B	C'
Ni(Co)-Mn(Cr,C)-In						
$Ni_{16}Mn_{12}In_4$	5.970	141.46	135.12	94.85	137.23	3.17
$Ni_{16}Mn_{11}Cr_1In_4$	5.970	140.45	134.50	95.18	136.48	2.97
$Ni_{16}Mn_{10}Cr_2In_4$	5.970	185.30	179.73	96.19	181.58	2.79
$Ni_{14}Co_2Mn_{12}In_4$	5.960	147.41	130.16	96.60	135.91	8.62
$Ni_{15}Co_1Mn_{11}Cr_1In_4$	5.964	144.33	133.68	97.13	137.23	5.33
$Ni_{14}Co_2Mn_{11}Cr_1In_4$	5.961	148.67	132.17	98.58	137.67	8.25
$Ni_{15}Co_1Mn_{10}Cr_2In_4$	5.962	183.54	173.66	98.44	176.95	4.94
$Ni_{14}Co_2Mn_{10}Cr_2In_4$	5.958	155.94	132.11	100.35	140.06	11.91
$Ni_{16}Mn_{11}C_1In_4$	5.970					-0.17
$Ni_{14}Co_2Mn_{11}C_1In_4$	5.931	148.71	135.81	92.99	140.11	6.45
Ni(Co)-Mn(Cr,C)-Sn						
$Ni_{16}Mn_{12}Sn_4$	5.951	163.97	139.78	98.38	147.84	12.10
$Ni_{16}Mn_{11}Cr_1Sn_4$	5.952	163.82	140.00	98.67	147.94	11.91
$Ni_{14}Co_2Mn_{12}Sn_4$	5.960	155.39	138.07	98.10	143.84	8.66
$Ni_{15}Co_1Mn_{11}Cr_1Sn_4$	5.948	162.24	141.63	100.16	148.50	10.31
$Ni_{14}Co_2Mn_{11}Cr_1Sn_4$	5.961	161.01	135.55	98.74	144.04	12.73
$Ni_{15}CoMn_{10}Cr_2Sn_4$	5.948	177.34	158.34	100.50	164.67	9.50
$Ni_{14}Co_2Mn_{10}Cr_2Sn_4$	5.957	160.81	136.00	99.74	144.27	12.41
$Ni_{16}Mn_{11}C_1Sn_4$	5.922	159.53	143.82	93.24	149.06	7.86
$Ni_{14}Co_2Mn_{11}C_1Sn_4$	5.933	156.88	138.59	93.03	144.69	9.14
$Ni_{13}Co_3Mn_{13}Sn_3$	5.937	163.16	134.48	98.57	144.04	14.34
$Ni_{13}Co_3Mn_{13}Sn_2Al_1$	5.898	165.49	137.30	101.93	146.69	14.10
*$Ni_2Mn_{1.6}Sn_{0.4}$		160.6	136.6	114.8	144.6	12.0
*$Ni_{1.9}Co_{0.1}Mn_{1.6}Sn_{0.4}$		163.5	136.9	116.2	145.8	13.3
*$Ni_{1.8}Co_{0.2}Mn_{1.6}Sn_{0.4}$		167.1	136.9	117.7	147.0	15.1

We also note that, according to the stability criterion for cubic crystals

$$C_{11} > |C_{12}|,\ C_{11} + 2C_{12} > 0,\ C_{44} > 0, \tag{3}$$

one of the investigated alloys ($Ni_{16}Mn_{11}C_1In_4$) has an unstable crystal lattice.

Similar calculations were made for the tetragonal phase of these alloys, the results are shown in Table 3. The distortions used for tetragonal structures and the corresponding strain energy densities are described in the last row of Table 1. The remaining alloys in the austenite phase are mechanically stable.

*Table 3. The equilibrium lattice constant [Å], the elastic constants (C_{11}, C_{12}, C_{44} [GPa]),
the bulk moduli (B [GPa]), and the shear modulus (C' [GPa]) of the investigated alloys
in the tetragonal phase (the results with asterisk (*) are obtained by the EMTO method in [11]).*

Name	a	c/a	C_{11}	C_{12}	C_{13}	C_{33}	C_{44}	C_{66}	B
Ni(Co)-Mn(Cr,C)-In									
$Ni_{16}Mn_{12}In_4$	5.970	1.3	343.49	144.47	174.49	173.11	83.21	39.88	173.08
$Ni_{16}Mn_{11}Cr_1In_4$	5.970	1.3	255.20	51.11	129.60	170.02	87.86	43.70	144.49
$Ni_{16}Mn_{10}Cr_2In_4$	5.970	1.3	233.75	30.34	155.23	235.47	76.65	34.54	122.62
$Ni_{14}Co_2Mn_{12}In_4$	5.960	1.3	228.17	46.72	127.14	168.26	90.77	51.01	135.38
$Ni_{15}Co_1Mn_{11}Cr_1In_4$	5.964	1.275	405.32	213.46	209.81	171.15	87.42	46.64	146.60
$Ni_{14}Co_2Mn_{11}Cr_1In_4$	5.961	1.275	311.88	120.14	161.97	166.85	87.77	47.37	166.45
$Ni_{15}Co_1Mn_{10}Cr_2In_4$	5.962	1.1	321.34	124.81	164.88	172.99	94.38	60.82	172.00
$Ni_{14}Co_2Mn_{10}Cr_2In_4$	5.958	1.2	231.44	65.56	128.93	172.05	97.30	61.78	142.39
$Ni_{14}Co_2Mn_{11}C_1In_4$	5.970	1.3	339.94	152.58	175.77	167.32	78.99	39.55	166.17
Ni(Co)-Mn(Cr,C)-Sn									
$Ni_{16}Mn_{12}Sn_4$	5.951	1.3	238.72	44.28	125.88	185.87	85.66	45.74	138.27
$Ni_{16}Mn_{11}Cr_1Sn_4$	5.952	1.3	194.62	29.18	178.80	314.43	75.10	35.48	46.72
$Ni_{14}Co_2Mn_{12}Sn_4$	5.960	1.3	238.26	36.89	125.68	179.48	83.51	44.74	135.42
$Ni_{15}Co_1Mn_{11}Cr_1Sn_4$	5.948	1.3	243.22	43.97	128.93	185.35	86.08	45.95	140.57
$Ni_{14}Co_2Mn_{11}Cr_1Sn_4$	5.961	1.3	239.52	36.38	126.76	178.99	83.88	44.83	135.97
$Ni_{15}Co_1Mn_{10}Cr_2Sn_4$	5.948	1.3	264.23	90.8	134.67	103.98	87.36	48.21	26.41
$Ni_{14}Co_2Mn_{10}Cr_2Sn_4$	5.957	1.3	182.96	45.55	108.93	105.65	91.30	49.83	100.38
$Ni_{16}Mn_{11}C_1Sn_4$	5.922	1.32	245.75	52.69	122.49	126.73	71.83	33.72	126.15
$Ni_{14}Co_2Mn_{11}C_1Sn_4$	5.933	1.3	229.25	37.18	120.34	178.44	78.01	39.50	136.95
$Ni_{13}Co_3Mn_{13}Sn_3$	5.937	1.25	242.15	43.86	123.17	184.78	94.75	50.14	139.25
*$Ni_2Mn_{1.6}Sn_{0.4}$			198.6	86.0	134.1	159.6	117.9	58.9	141.4
*$Ni_{1.9}Co_{0.1}Mn_{1.6}Sn_{0.4}$			197.9	89.5	134.0	164.1	117.6	65.1	142.6
*$Ni_{1.8}Co_{0.2}Mn_{1.6}Sn_{0.4}$			199.8	90.6	134.2	168.4	122.1	69.3	144.0

The elastic stiffness coefficients for martensitic phases shown in Table 3 obey the generalized elastic stability criteria [12] for tetragonal crystals. This indicates that L_{21} structures are mechanically stable.

However, single crystals are not used in practical applications. Moreover, single-crystal samples are often not used in experiments on the determination of elastic moduli, hence individual values of the constants C_{ij} cannot be obtained. Therefore, it is necessary to consider the properties of polycrystalline samples based on previously determined properties of single crystals.

To describe the elastic properties of a polycrystal, two constants of elasticity are sufficient: bulk modulus B and shear modulus G. These values can be obtained using the Reuss-Voigt-Hill averaging method [13-15]. The Voigt and Reiss methods represent the upper and lower boundaries, respectively, to the isotropic elastic modulus. The results of the calculation for bulk modulus B and shear modulus G for austenite and martensitic structures are given in Table 4.

Shape Memory Alloys – SMA 2018   Materials Research Forum LLC
Materials Research Proceedings **9** (2018) 83-91   doi: http://dx.doi.org/10.21741/9781644900017-16

Table 4. Volumetric moduli (B), shear moduli (G_H) and their ratio for the investigated alloys.

Name	B_H [GPa]		G_H [GPa]		B/G	
	cube	tetragonal	cube	tetragonal	cube	tetragonal
Ni(Co)-Mn(Cr, C)-In						
$Ni_{16}Mn_{12}In_4$	137.2	189.1	32.9	51.8	4.2	3.7
$Ni_{16}Mn_{11}Cr_1In_4$	136.5	144.5	32.7	55.6	4.2	2.6
$Ni_{16}Mn_{10}Cr_2In_4$	181.6	138.2	32.8	45.5	5.5	3.0
$Ni_{14}Co_2Mn_{12}In_4$	135.9	135.8	40.2	57.7	3.4	2.4
$Ni_{15}Co_1Mn_{11}Cr_1In_4$	137.2	198.2	36.4	52.1	3.8	3.8
$Ni_{14}Co_2Mn_{11}Cr_1In_4$	137.7	176.5	40.4	55.7	3.4	3.2
$Ni_{15}Co_1Mn_{10}Cr_2In_4$	177.0	181.8	36.3	64.8	4.9	2.8
$Ni_{14}Co_2Mn_{10}Cr_2In_4$	140.1	142.4	45.1	66.4	3.1	2.1
$Ni_{14}Co_2Mn_{11}C_1In_4$	140.1	186.2	36.5	49.4	3.8	3.8
Ni(Co)-Mn(Cr, C)-Sn						
$Ni_{16}Mn_{12}Sn_4$	147.8	138.9	44.7	57.5	3.3	2.4
$Ni_{16}Mn_{11}Cr_1Sn_4$	147.9	105.43	44.6	40.6	3.3	2.6
$Ni_{14}Co_2Mn_{12}Sn_4$	143.8	136.19	40.7	55.08	3.5	2.5
$Ni_{15}Co_1Mn_{11}Cr_1Sn_4$	148.5	141.14	43.3	57.18	3.4	2.5
$Ni_{14}Co_2Mn_{11}Cr_1Sn_4$	144.0	136.75	45.5	54.92	3.2	2.5
$Ni_{15}Co_1Mn_{10}Cr_2Sn_4$	164.7	88.35	42.5	30.78	3.9	2.9
$Ni_{14}Co_2Mn_{10}Cr_2Sn_4$	144.3	105.6	45.5	31.1	3.2	3.3
$Ni_{16}Mn_{11}C_1Sn_4$	149.1	130.5	38.3	40.1	3.9	3.2
$Ni_{14}Co_2Mn_{11}C_1Sn_4$	144.7	123.18	39.7	51.33	3.6	2.4
$Ni_{13}Co_3Mn_{13}Sn_3$	144.0	139.2	47.2	63.2	3.1	2.2

The ratio of the bulk modulus of elasticity to the shear modulus for polycrystalline samples *(B/G)* proposed by Pugh [16], characterizes the plasticity of the material. A high *(B/G)* value is associated with better ductility, while a lower value corresponds to a more brittle material. The critical value that distinguishes plastic material from brittle material is about 1.75. According to our calculations, the *B/G* value is approximately 4 for the cubic phase of the compound Ni(Co)-Mn(Cr, C)-In and about 3.4 for Ni(Co)-Mn(Cr, C)-Sn. For the tetragonal phase Ni(Co)-Mn(Cr, C)-In, this value is 3.1, and for Ni(Co)-Mn(Cr, C)-Sn is 2.4. It can be seen that in the cubic phase the compounds have a greater plasticity than in the tetragonal phase.

At the final stage, the values of such elastic characteristics as the Young's modulus (*E*), the Poisson's ratio (*ν*) were determined using the known relationships between the elastic moduli presented in Tables 2 and 3. The obtained values of the elastic moduli were used to calculate the Debye temperature [16]

$$\Theta_D = \hbar(6\pi^2 V^{1/2} n)^{1/3} f(\nu)\sqrt{\frac{B}{k^2 M}} N_A, \tag{4}$$

where *f* is an auxiliary function that depends only on the Poisson ratio

$$f(\nu) = \left\{ 3\left[2\left(\frac{2+2\nu}{3-6\nu}\right)^{3/2} + \left(\frac{1+\nu}{3-3\nu}\right)^{3/2}\right]^{-1}\right\}^{1/3}. \tag{5}$$

Shape Memory Alloys – SMA 2018
Materials Research Proceedings **9** (2018) 83-91

Materials Research Forum LLC
doi: http://dx.doi.org/10.21741/9781644900017-16

The results of these calculations are given in Table 5.

Table 5. Poisson's ratio, Young's modulus and the Debye temperature for the investigated alloys (the results with asterisk (*) are obtained by the EMTO method in [11]).

Name	v		E [GPa]		Θ [K]	
	cube	tetragonal	cube	tetragonal	cube	tetragonal
Ni(Co)-Mn(Cr, C)-In						
$Ni_{16}Mn_{12}In_4$	0.39	0.37	91.29	142.44	289.95	363.36
$Ni_{16}Mn_{11}Cr_1In_4$	0.39	0.33	90.84	130.19	289.46	354.79
$Ni_{16}Mn_{10}Cr_2In_4$	0.41	0.35	92.69	123.03	290.98	339.86
$Ni_{14}Co_2Mn_{12}In_4$	0.37	0.31	109.81	151.54	319.65	379.79
$Ni_{15}Co_1Mn_{11}Cr_1In_4$	0.38	0.38	100.21	143.59	304.57	364.50
$Ni_{14}Co_2Mn_{11}Cr_1In_4$	0.37	0.36	110.37	151.24	320.38	375.82
$Ni_{15}Co_1Mn_{10}Cr_2In_4$	0.40	0.34	101.84	173.71	305.48	404.85
$Ni_{14}Co_2Mn_{10}Cr_2In_4$	0.35	0.30	122.26	172.48	338.27	407.05
$Ni_{14}Co_2Mn_{11}C_1In_4$	0.38	0.38	100.72	136.12	307.35	357.47
Ni(Co)-Mn(Cr, C)-Sn						
$Ni_{16}Mn_{12}Sn_4$	0.36	0.32	121.83	151.60	335.14 / 332.2*	377.81 / 356.0*
$Ni_{16}Mn_{11}Cr_1Sn_4$	0.36	0.33	121.56	106.67	335.00 / 350.8*	316.15 / 377.0*
$Ni_{14}Co_2Mn_{12}Sn_4$	0.37	0.32	111.64	145.62	320.44	370.19
$Ni_{15}Co_1Mn_{11}Cr_1Sn_4$	0.37	0.32	118.33	141.14	330.08	377.05
$Ni_{14}Co_2Mn_{11}Cr_1Sn_4$	0.36	0.32	123.49	145.31	338.33	370.0
$Ni_{15}Co_1Mn_{10}Cr_2Sn_4$	0.38	0.34	117.27	82.73	327.81	277.67
$Ni_{14}Co_2Mn_{10}Cr_2Sn_4$	0.36	0.36	123.46	85.11	338.39	281.76
$Ni_{16}Mn_{11}C_1Sn_4$	0.38	0.36	105.74	109.22	313.40	320.63
$Ni_{14}Co_2Mn_{11}C_1Sn_4$	0.37	0.33	109.12	165.05	319.15	369.34
$Ni_{13}Co_3Mn_{13}Sn_3$	0.35	0.33	127.54	177.66	348.65	406.56

It is seen from this table that when passing from a cubic lattice to a tetragonal lattice, the Poisson's coefficients decrease, and the Young's modulus increases, this also confirms that in the tetragonal phase the compounds under study have a lower plasticity than in the cubic phase. Moreover, the Debye temperature calculated by us is in good agreement with the values calculated by other authors by the EMTO method [11]. Also, as can be seen from the Table 5, the Debye temperature for austenitic phase is less than for the martensitic phase, which shows that the specific heat of C_P for austenite is larger. These results agree with similar results for the Ni_2MnGa alloy [18-19].

Conclusion

In this work, we introduced first-principles calculation to study structural and elastic properties of Ni(Co)-Mn(Cr, C)-In and Ni(Co)-Mn(Cr, C)-Sn alloys. Our calculated components of the tensor of elastic constants are in good agreement with the results of other authors. The discrepancy for the bulk module is about 2 %.

In this paper we show that the majority (except for $Ni_{16}Mn_{11}C_1In_4$) of the investigated alloys have a mechanically stable crystal lattice. We also note that the small value of the tetragonal shift constant for the austenite phase indicates that the alloys are close to a phase transition. In

Shape Memory Alloys – SMA 2018
Materials Research Proceedings **9** (2018) 83-91

Materials Research Forum LLC
doi: http://dx.doi.org/10.21741/9781644900017-16

addition, the elastic constants of alloys, such as the Poisson's ratio, the Young's modulus and the Debye temperature, are determined. These properties are important for practical applications. In addition, the plasticity of these alloys was evaluated.

Acknowledgements
This work was supported by Russian Science Foundation No. 17-72-20022.

References

[1] A.N. Vasiliev, V.D. Buchelnikov, et al., Shape memory ferromagnets, Phys.-Usp. 46 (2003) 559-689. https://doi.org/10.1070/PU2003v046n06ABEH001339

[2] P. Entel et al., Fundamental Aspects of Magnetic Shape Memory Alloys: Insights from Ab Initio and Monte Carlo Studies, Materials Science Forum 635 (2010) 3-12. https://doi.org/10.4028/www.scientific.net/MSF.635.3

[3] L. Manosa, A. Gonzalez-Comas, E. Obrado, A. Planes, Anomalies related to the TA2-phonon-mode condensation in the Heusler Ni2MnGa alloy, Phys. Rev. B 55 (1997) 11068-11071. https://doi.org/10.1103/PhysRevB.55.11068

[4] M. Stipcich, L. Manosa, A. Planes, M. Morin, J. Zarestky, T. Lograsso, C. Stassis, Elastic constants of Ni−Mn−Ga magnetic shape memory alloys, Phys. Rev. B 70 (2004) 054115. https://doi.org/10.1103/PhysRevB.70.054115

[5] P.J. Brown, J. Crangle, T. Kanomata, M. Matsumota, K.-U. Neumann, B. Ouladdiaf, K.R.A. Ziebeck, The crystal structure and phase transitions of the magnetic shape memory compound Ni_2MnGa, J. Phys.: Condens. Matter. 14 (2002) 10159-10171. https://doi.org/10.1088/0953-8984/14/43/313

[6] Q.M. Hu, C.M. Li, R. Yang, S.E. Kulkova, D.I. Bazhanov, B. Johansson, L. Vitos, Site occupancy, magnetic moments, and elastic constants of off-stoichiometric Ni_2MnGa from first-principles calculations, Phys. Rev. B 79 (2009) 144-112. https://doi.org/10.1103/PhysRevB.79.144112

[7] S. Ozdemir Kart, T.Cagın, Elastic properties of Ni_2MnGa from first-principles calculations, J. Alloy. Comp. 508 (2010) 177-183. https://doi.org/10.1016/j.jallcom.2010.08.039

[8] G. Kresse, D. Joubert, From ultrasoft pseudopotentials to the projector augmented-wave method, Phys. Rev. B. 59 (1999) 1758-1775. https://doi.org/10.1103/PhysRevB.59.1758

[9] H.J. Monkhorst, J.D. Pack, Special points for Brillouin-zone integrations, Phys. Rev. 13 (1976) 5188. https://doi.org/10.1103/PhysRevB.13.5188

[10] J.P. Perdew, K. Burke, M. Ernzerhof, Generalized Gradient Approximation Made Simple, Phys. Rev. Lett. 77 (1996) 3865-3868. https://doi.org/10.1103/PhysRevLett.77.3865

[11] L. Chun-Mei, H. Qing-Miao, Y. Rui, J. Borje, V. Levente, Theoretical investigation of the magnetic and structural transitions of Ni-Co-Mn-Sn metamagnetic shape-memory alloys, Phys. Rev. B. 92 (2015) 024105.

[12] I.A. Abrikosov, A.Y. Nikonov, A.V. Ponomareva, A.I. Dmitriev, S.A. Barannikova, Theoretical modeling of thermodynamic and mechanical properties of the pure components of Ti and Zr based alloys using the exact muffin-tin orbitals method, Usp. Fiz. Met. 14 (2013) 319-352. https://doi.org/10.15407/ufm.14.04.319

[13] A. Reuss Berechung der Fliebgrenze von Mischkristallen auf Grund der Plastizitats-bedingung fur Einkristalle, Zs. Angew. Math. und Mech., 1 (1929) 49-58.

Shape Memory Alloys – SMA 2018   Materials Research Forum LLC
Materials Research Proceedings **9** (2018) 83-91   doi: http://dx.doi.org/10.21741/9781644900017-16

[14] W. Voight Lehrbuch der Kristallphysik, Berlin, Teubner 1928.

[15] R. Hill, The Elastic Behavior of a Crystalline Aggregate, Proc. Phys. Soc. A 65 (1952) 349-354. https://doi.org/10.1088/0370-1298/65/5/307

[16] S.F. Pugh, Relations between the elastic moduli and the plastic properties of polycrystalline pure metals, Phil. Mag. 45 (1954) 823-843. https://doi.org/10.1080/14786440808520496

[17] V. Levente, Computational Quantum Mechanics for Materials Engineers, Springer, Sweden, 2007.

[18] M. Kreissl, K.-U. Neumann, T. Stephans, K.R.A. Ziebeck, The influence of atomic order on the magnetic and structural properties of the ferromagnetic shape memory compound Ni_2MnGa, J. Phys.: Condens. Matter 15 (2003) 3831-3839. https://doi.org/10.1088/0953-8984/15/22/317

[19] E. Cesari, V.A. Chernenko, J. Font, J. Muntasell, ac technique applied to c_p measurements in Ni–Mn–Ga alloys, Thermochim. Acta 433 (2005) 153-156. https://doi.org/10.1016/j.tca.2005.02.029

Shape Memory Alloys – SMA 2018
Materials Research Proceedings 9 (2018) 92-97

Materials Research Forum LLC
doi: http://dx.doi.org/10.21741/9781644900017-17

Properties of Fe-Ga and Fe-Ga-V Alloys: *Ab Initio* Study

Mariya V. Matyunina[1,a *], Mikhail A. Zagrebin[1,2,b], Vladimir V. Sokolovskiy[1,2,c], Vasiliy D. Buchelnikov[1,2,d]

[1]Chelyabinsk State University, 129 Brat'ev Kashirinykh Str., Chelyabinsk 454001, Russia

[2]National University of Science and Technology 'MISiS', 4 Leninskiy Prospect, Moscow 119991, Russia

[a]matunins.fam@mail.ru, [b]miczag@mail.ru, [c]vsokolovsky84@mail.ru, [d]buche@csu.ru

*corresponding author

Keywords: *Ab Initio* Calculations, Shear Modulus, Curie Temperature, Fe-Ga, Fe-Ga-V

Abstract. Structural and magnetic properties of $Fe_{73.44}Ga_{26.56}$ and $Fe_{73.44}Ga_{17.18}V_{9.38}$ alloys have been studied by means of *ab initio* calculations. It was shown that the adding of V atoms into Fe-Ga alloy stabilized the $D0_3$ structure and changes the shear modulus sign from negative to positive. Moreover, the total magnetic moment is decreased by 10 % and 16 % for $D0_3$ and $L1_2$ structures, respectively. The Curie temperatures of the studied compositions were estimated. It was found that in case of $Fe_{73.44}Ga_{26.56}$ the calculated Curie temperature of $L1_2$ structure is in an agreement with the experimental data.

Introduction

The Fe-Ga alloys are well-known materials due to their giant saturation magnetostriction at the low magnetic field, which makes them attractive for the potential application in magnetostrictive drives and sensors. Moreover, ferromagnetic Fe-Ga alloys exhibit both deformations induced by a magnetic field and a shape memory effect [1]. Restorff et al. [2] have been shown the influence of adding elements to Fe-Ga alloys on the tetragonal magnetostriction constant and magnetoelastic coupling. Adding 15 at.% Al to the Fe-Ga alloys allows to achieve a magnetostriction of 200 ppm with only 5 at.% Ga content. Nevertheless, the two peaks curve of magnetostriction the Fe-Ga-Al alloy at low Al content become the single magnetostriction peak curve at high Al concentrations. The addition of ≈ 7 at.% of Ge atoms to Fe-Ga system with Ga content more than 15 at.% change sign of magnetostriction from positive to negative. In the experimental work [1] the tetragonal magnetostriction and phase transformation of the $Fe_{73}Ga_{18}V_9$ alloy was studied. It was shown the adding of V atoms to $Fe_{73}Ga_{27}$ compound does not result in a higher magnetocrystalline anisotropy constant and saturation magnetization. Moreover, the $D0_3 \rightarrow L1_0$-like tetragonal martensite$\rightarrow L1_2$ phase transformation was not observed in Fe-Ge, in contrast to $Fe_{73}Ga_{27}$.

The present work is aimed the complex study of the influence of V-doping on the magnetic and structural properties of $Fe_{73}Ga_{27}$ alloy by zero-temperature *ab initio* investigation and mean field approximation.

Computation details

At the first step of our calculations, the geometric optimization and calculations of elastic properties of crystal structures were conducted by density functional theory implemented in Vienna *Ab initio* Simulation Package (VASP) [3,4] with an account of supercell approach. Generalized gradient approximation (GGA) in Perdew-Burke-Ernzerhof (PBE) form [5] were used. The *k*-point mesh was generated using the Monkhorst-Pack scheme [6] with a grid of $8 \times 8 \times 8$ points. The interaction between ions and electrons was described with the projector-augmented wave method (PAW) [3,4]. Following configurations of the PAW-potentials were

Published under license by Materials Research Forum LLC.

Shape Memory Alloys – SMA 2018
Materials Research Proceedings **9** (2018) 92-97

Materials Research Forum LLC
doi: http://dx.doi.org/10.21741/9781644900017-17

selected: Fe($3p^6 3d^7 4s^1$), Ga($3d^{10} 4s^2 4p^1$), and V($3d^{10} 4s^2 4p^1$). The value of the kinetic energy cut-off for the augmentation charges was 800 eV, while the plane-wave cut-off energy was 400 eV. All calculations were performed for $T = 0$ K.

Fe$_{73.44}$Ga$_{26.56}$ and Fe$_{73.44}$Ga$_{17.18}$V$_{9.38}$ compositions were considered in the present study. Calculations were conducted for 64-atom supercells, in terms of which studied compositions are Fe$_{47}$Ga$_{17}$ and Fe$_{47}$Ga$_{11}$V$_6$, correspondingly. Two types of phases were studied for both compositions. They are D0$_3$ with the BiF$_3$-type structure (*Fm-3m*, #225) and the L1$_2$ with a Cu$_3$Au-type structure (*Pm-3m*, #221) (see Fig. 1). In the case of L1$_2$ structure the atoms occupy the following positions: $1a$ (0, 0, 0) – Ga$_1$ or Ga$_1$ + V and $3c$ (0, 0.5, 0.5) – Fe, Ga$_2$, and for the D0$_3$ structure the atoms occupy the $4a$ (0, 0, 0) – Ga$_1$ or Ga$_1$ + V, $4b$ (0.5, 0.5, 0.5) – Fe$_2$, Ga$_2$ and $8c$ (0.25, 0.25, 0.25) – Fe$_1$. To create the non-stoichiometric compositions, 17 atoms of Fe were replaced by Ga in case of Fe$_{47}$Ga$_{17}$, while for 11 atoms of Ga and 6 atoms of V for the second one.

The elastic properties calculations for cubic structures were performed according to the works of Abrikosov et al. [7], Kart and Cagin [8]. The elastic constants were calculated by determining the total energies of the alloy obtained for a number of small deformations with volume conservation (V = const) [7]. For calculation of the bulk modulus B and two shear moduli C' and C_{44}, we used the isotropic, orthorhombic, and monoclinic deformations of strain tensors, respectively. The distortion parameter was taken in the range from –0.03 to 0.03 with the step 0.01.

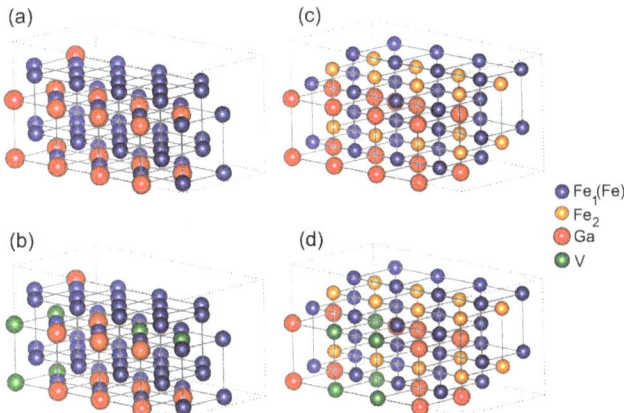

Fig. 1. *The 64-atom supercells of (a, b) L1$_2$ and (c, d) D0$_3$ crystal structures for Fe$_{73.44}$Ga$_{26.56}$ and Fe$_{73.44}$Ga$_{17.18}$V$_{9.38}$ alloys, respectively.*

At the second step, using the optimized lattice parameters and the expression proposed by Liechtenstein et al. [9], the magnetic exchange coupling constants, J_{ij} were obtained. The calculations were performed by means of the SPR-KKR (spin-polarized relativistic Korringa-Kohn-Rostoker) package [10] with an account of the local density approximation in Vosko-Wilk-Nusair parametrization (VWN) [11] and coherent potential approximation (CPA). All calculations converged to 0.01 mRy of total energy.

Finally, obtained with the *ab initio* calculations magnetic exchange interactions the Curie temperature was calculated by means of the mean field approximation (MFA). Using the well-established Heisenberg model in the framework of MFA, the Curie temperature (T_C) was obtained by solving the system of coupled equations [12]

$$\left\langle s^{\mu} \right\rangle = \frac{2}{3k_{\mathrm{B}}T} \sum_{\upsilon} J_0^{\mu\upsilon} \left\langle s^{\upsilon} \right\rangle, \tag{1}$$

where k_{B} is Boltzmann constant, $J_0^{\mu\upsilon} = \sum_{i,j} J_{ij}$ is the effective exchange parameter; μ and υ are

the different sublattices; $\left\langle s^{\upsilon} \right\rangle$ is the average z component of an atom spin located in the υ sublattice.

Results and Discussion
In this section, we present the results of *ab initio* investigations of the structural and magnetic properties of Fe-Ga (binary) and Fe-Ga-V (ternary) alloys. At the first step, the geometric optimization and calculations of elastic constants based on equilibrium lattice parameters were done for D0$_3$ and L1$_2$ structures of Fe$_{73.44}$Ga$_{26.56}$ and Fe$_{73.44}$Ga$_{17.18}$V$_{9.38}$ alloys. Calculated equilibrium lattice parameters a_0 (corresponding to a minimum value of energy E_0), ground state energy E_0, shear moduli C' and C_{44}, bulk B and Young's E moduli for D0$_3$ and L1$_2$ cubic structures of Fe-Ga and Fe-Ga-V alloys are presented in Table 1.

Table 1. The calculated equilibrium lattice parameters a_0 (Å), ground state energy E_0 (meV), shear moduli C' and C_{44}, bulk B and Young's E moduli (in GPa), for D0$_3$, and L1$_2$ cubic structures of Fe$_{73.44}$Ga$_{26.56}$ and Fe$_{73.44}$Ga$_{17.18}$V$_{9.38}$ alloys. The negative value of C' is bolding.

Alloy	Phase	a_0	E_0	C'	C_{44}	B	E
Fe$_{73.44}$Ga$_{26.56}$	D0$_3$	5.78	−6920	**−1.44**	127	138	99
	L1$_2$	3.66	−6959	33.35	118	155	186
Fe$_{73.44}$Ga$_{17.18}$V$_{9.38}$	D0$_3$	5.76	−7518	21.04	132	137	168
	L1$_2$	3.64	−7506	7.91	91	155	105

The lattice parameters of the Fe-Ga alloys are higher than for Fe-Ga-V ones by 0.3 % and 0.5 %, for D0$_3$ and L1$_2$ structures, respectively. The knowledge of lattice parameters of considered compositions allows us to perform the calculations of elastic constants. The shear C', bulk B, and Young's E moduli are defined by next expressions:

$$C' = \frac{\left(C_{11} - C_{12}\right)}{2}, \; B = \frac{C_{11} + 2C_{12}}{3}, \text{ and } E = \frac{9BG}{3B + G}, \tag{2}$$

where G is the Hill average shear modulus [10], which is defined as

$$G = \frac{1}{2}\left(\frac{2C' + 3C_{44}}{5} + \frac{5C'C_{44}}{2C_{44} + 3C'} \right). \tag{3}$$

As can be seen from Table 1 for binary alloy, the L1$_2$ phase is stable ($E_0^{\mathrm{L1_2}} > E_0^{\mathrm{D0_3}}$), and the negative shear modulus C' indicates the instability of D0$_3$ phase. However, adding of V atoms stabilizes the D0$_3$ structure ($E_0^{\mathrm{D0_3}} > E_0^{\mathrm{L1_2}}$), and C' is positive for both structures.

To investigate the possibility of martensitic transformation in these alloys, the calculations of the total energy as a function of the tetragonal distortion (c/a) of the cubic structure were performed. The volume $V_0 \approx a_0^3 \approx a^2 c$ does not change with tetragonal distortions. Fig. 2 presents the total energy differences per atom with respect to an equilibrium value of the favorable phase ($c/a = 1$) for Fe$_{73.44}$Ga$_{26.56}$ ($E_0^{\mathrm{L1_2}}$) and Fe$_{73.44}$Ga$_{17.18}$V$_{9.38}$ ($E_0^{\mathrm{D0_3}}$) alloys as functions of tetragonal distortion (c/a). It is seen that there is no martensitic transition in Fe-Ga alloys with the L1$_2$

Shape Memory Alloys – SMA 2018   Materials Research Forum LLC
Materials Research Proceedings **9** (2018) 92-97   doi: http://dx.doi.org/10.21741/9781644900017-17

structure. For metastable $D0_3$ structure, the cubic phase is not stable. For this structure the tetragonal state with $c/a = 1.25$ is more favorable. In the case of Fe-Ga-V alloy with the $L1_2$ metastable structure, the martensitic transition is possible in the tetragonal state with $c/a = 0.8$. The transition from the $D0_3$ cubic phase to the tetragonal phase is not possible. However, it should be noted that the energy difference between $D0_3$ and $L1_2$ structures is small for this alloy. It indicates that a change in the vanadium content can lead to a martensitic transition from the $D0_3$ cubic phase to the tetragonal phase corresponding to $L1_0$.

The recent investigation of the aged $Fe_{73}Ga_{27}$ alloy using the transmission electron microscopy [1] shows the $L1_2$ structures precipitate through the $D0_3$ nanoclusters transformations into an intermediate tetragonal phase ($L1_0$-like martensite +A2) via Bain distortion. However, the investigation of V-doped $Fe_{73}Ga_{27}$ alloy found that the $D0_3$ phase is an ordered structure and the phase transformation of $D0_3 \rightarrow$ tetragonal martensite $\rightarrow L1_2$ was not found. So, our calculations did not confirm experimental results for $Fe_{73}Ga_{27}$ alloy. It can be explained the conditions of the preparations of the samples.

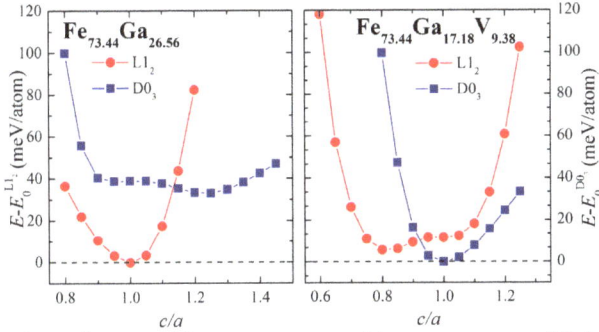

Fig. 2. The total energy differences per atom with respect to an equilibrium value of the favorable phase for $Fe_{73.44}Ga_{26.56}$ ($E_0^{L1_2}$) and $Fe_{73.44}Ga_{17.18}V_{9.38}$ ($E_0^{D0_3}$) alloys as functions of tetragonal distortion (c/a).

Calculated exchange interaction parameters J_{ij} as a function of distance (d/a) between atoms i and j for $Fe_{73.44}Ga_{26.56}$ and $Fe_{73.44}Ga_{17.18}V_{9.38}$ alloys with (a, b) $L1_2$ and (c, d) $D0_3$ structures at their equilibrium lattice parameters are shown in Fig. 3. The largest ferromagnetic (FM) interactions ($J_{ij} > 0$) are observed in the first coordination shell between nearest Fe atoms (Fe_1-Fe_2) located at different sublattices of $D0_3$ structures. It is clearly seen, that the adding of V atoms to Fe-Ga alloy enhance this interaction, however, in the case of the $L1_2$ structure, the interaction between nearest Fe-Fe atoms is weaker. In case of $L1_2$ phase, the split of the Fe-Fe exchange parameters into two FM contributions within the second coordination shell ($d/a = 1$) is observed for both compositions. In this instance, each Fe atom has six nearest neighbors at the distance of a_0 and interacts ferromagnetically with two Fe atoms providing $J_{ij} \approx 7.59/10.71$ meV and with four Fe atoms located in (x, y) plane and providing $J_{ij} \approx 2.86/2.80$ meV for Fe-Ga/Fe-Ga-V alloys, respectively.

Calculated magnetic moment of Fe atoms and total magnetic moments for $D0_3$ and $L1_2$ structures of $Fe_{73.44}Ga_{26.56}$ and $Fe_{73.44}Ga_{17.18}V_{9.38}$ alloys are presented in Table 2. We would like to note, that calculated atomic moments of Ga do not depend on V-doping for both alloys. Nevertheless, for the $D0_3$ and $L1_2$ structures magnetic moment of Fe_1, Fe_2, and Fe atoms change by 6 %, 2 %, and 9 %, respectively. The total magnetic moment decreases by 10 % and 16 % with vanadium adding for $D0_3$ and $L1_2$ structure, respectively.

Shape Memory Alloys – SMA 2018   Materials Research Forum LLC
Materials Research Proceedings **9** (2018) 92-97   doi: http://dx.doi.org/10.21741/9781644900017-17

Table 2. The calculated magnetic moments of Fe atoms (in μ_B) and total magnetic moments (in μ_B/f.u.), Curie temperature T_C^{calc} (in K) for $D0_3$ and $L1_2$ cubic structures of $Fe_{73.44}Ga_{26.56}$ and $Fe_{73.44}Ga_{17.18}V_{9.38}$ alloys in comparison with available experimental data [13] denoted T_C^{exp} (in K).

Phase	$D0_3$					$L1_2$			
Alloy	μ_{Fe_1}	μ_{Fe_2}	μ_{tot}	T_C^{calc}	T_C^{exp}	μ_{Fe}	μ_{tot}	T_C^{calc}	T_C^{exp}
$Fe_{73.44}Ga_{26.56}$	1.76	2.46	5.73	1094	675	2.29	6.57	1203	1015
$Fe_{73.44}Ga_{17.18}V_{9.38}$	1.65	2.41	5.15	1325	-	2.08	5.54	1008	-

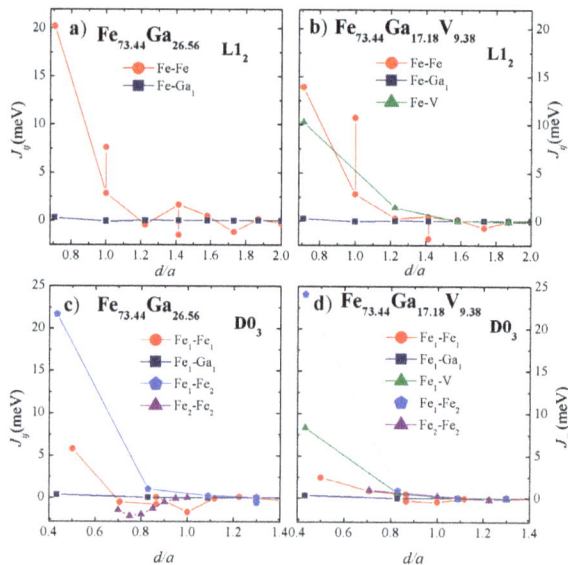

Fig. 3. The calculated exchange interaction parameters J_{ij} as a function of distance (d/a) between atoms i and j for $Fe_{73.44}Ga_{26.56}$ and $Fe_{73.44}Ga_{17.18}V_{9.38}$ alloys with (a, b) $L1_2$ and (c, d) $D0_3$ structures.

Finally, the Curie temperature was defined using the MFA (See Table 2). As can be seen, the Curie temperature is higher for stable phases $L1_2$ for Fe-Ga and $D0_3$ for Fe-Ga-V alloys. The T_C calculations for Fe-Ga alloy with stable $L1_2$ phase is in agreement with the experimental data, in contrast to $D0_3$ structure, the Curie temperature of which is significantly higher.

Summary

In this work the structural and magnetic properties of $Fe_{73.44}Ga_{26.56}$ and $Fe_{73.44}Ga_{17.18}V_{9.38}$ alloys have been studied by means of *ab initio* calculations. The elastic constants, exchange interaction parameters J_{ij}, and Curie temperature were calculated. It was shown that the adding V atoms to Fe-Ga alloys stabilized the $D0_3$ structure. Moreover, the shear modulus C' of the $D0_3$ phase changes sign from negative to positive. The total magnetic moment and magnetic moments of Fe atoms decrease for both structures. The investigation of the total energy differences of Fe-Ga and Fe-Ga-V alloys as a function of the tetragonal distortion (c/a) showed that the martensitic transformation in these alloys is not observed. Calculations of Curie temperatures shown that the

Curie temperature of the stable $L1_2$ structure of Fe-Ga is in accordance with experimental value, while it is significantly higher than the experimental one for $D0_3$ structure.

Acknowledgments
This work was supported by Russian Science Foundation No. 18-12-00283.

References

[1] Y.C. Lin, C.F. Lin, Microstructures and Magnetic Properties of Fe-Ga and Fe-Ga-V Ferromagnetic Shape Memory Alloys, IEEE T. Mag. 51 (2015) 2505204. https://doi.org/10.1109/INTMAG.2015.7157539

[2] J. B. Restorff,1 M. Wun-Fogle, K. B. Hathaway, A. E. Clark, T. A. Lograsso, G. Petculescu, Tetragonal magnetostriction and magnetoelastic coupling in Fe-Al, Fe-Ga, Fe-Ge, Fe-Si, Fe-Ga-Al, and Fe-Ga-Ge alloys, J. Appl. Phys. 111 (2012), 023905. https://doi.org/10.1063/1.3674318

[3] G. Kresse, J. Furthmüller, Efficient iterative schemes for ab initio total-energy calculations using a plane-wave basis set, Phys. Rev. B 54 (1996) 11169-11186. https://doi.org/10.1103/PhysRevB.54.11169

[4] G. Kresse, D. Joubert, From ultrasoft pseudopotentials to the projector augmented-wave method, Phys. Rev. B 59 (1999) 1758-1775. https://doi.org/10.1103/PhysRevB.59.1758

[5] J. P. Perdew, K. Burke, M. Enzerhof, Generalized Gradient Approximation Made Simple, Phys. Rev. Lett. 77 (1996) 3865-3868. https://doi.org/10.1103/PhysRevLett.77.3865

[6] H.J. Monkhorst, J.D. Pack, Special points for Brillouin-zone integrations, Phys. Rev. B. 13 (1976) 5188-5192. https://doi.org/10.1103/PhysRevB.13.5188

[7] I.A. Abrikosov, A.Y. Nikonov, A.V. Ponomareva, A.I. Dmitriev, S.A. Baran-nikova, Theoretical modeling of thermodynamic and mechanical properties of the pure components of Ti and Zr based alloys using the exact muffin-tin orbitals method, Usp. Fiz. Met. 14 (2013) 319-352. https://doi.org/10.15407/ufm.14.04.319

[8] S.O. Kart, T.Cagın, Elastic properties of Ni_2MnGa from first-principles calculations, J. Alloy. Comp. 508 (2010) 177-183. https://doi.org/10.1016/j.jallcom.2010.08.039

[9] A.I. Liechtenstein, M.I. Katsnelson, V.P. Antropov, V.A. Gubanov, Local spin density functional approach to the theory of exchange interactions in ferromagnetic metals and alloys, J. Magn. Magn. Mater. 67 (1987) 65-74. https://doi.org/10.1016/0304-8853(87)90721-9

[10] H. Ebert, D. Ködderitzsch and J. Minár, Calculating condensed matter properties using the KKR-Green's function method-recent developments and applications, Rep. Prog. Phys. 74 (2011) 096501. https://doi.org/10.1088/0034-4885/74/9/096501

[11] S.H. Vosko, L. Wilk, M. Nusair, Accurate spin-dependent electron liquid correlation energies for local spin density calculations: a critical analysis, Canadian J. of Phys., 58 (1980) 1200-1211. https://doi.org/10.1139/p80-159

[12] V.V. Sokolovskiy, V.D. Buchelnikov, M.A. Zagrebin, S.V. Taskaev, V.V. Khovaylo and P. Entel, *Ab initio* study of magnetic properties of Fe-Mn-Al Heusler alloys, Mater. Res. Soc. Symp. Proc. 1581 (2013) 44-49. https://doi.org/10.1557/opl.2013.888

[13] H. Okamoto, The Fe-Ga (Iron-Gallium) System, Bull. Alloy Phase Diagrams 11 (1990) 576-581. https://doi.org/10.1007/BF02841721

Shape Memory Alloys – SMA 2018
Materials Research Proceedings 9 (2018) 98-103

Materials Research Forum LLC
doi: http://dx.doi.org/10.21741/9781644900017-18

Phase Transitions in Ni(Co)-Mn-Sn Heusler Alloys: First-Principles Study

Vladimir V. Sokolovskiy[1,2,a], Olga N. Miroshkina[1,b*], Mikhail A. Zagrebin[1,3,c], Vasiliy D. Buchelnikov[1,2,d]

[1]Chelyabinsk State University, 129 Brat'ev Kashirinykh Str., Chelyabinsk 454001, Russia

[2]National University of Science and Technology "MISiS", 4 Leninskiy Prospect, Moscow 119991, Russia

[3]National Research South Ural State University, 76 Lenina Prospect, Chelyabinsk 454080, Russia

[a]vsokolovsky84@mail.ru, [b]miroshkina.on@yandex.ru, [c]miczag@mail.ru, [d]buche@csu.ru

*corresponding author

Keywords: *Ab Initio* Calculations, Heusler Alloys, First-Principles Study, Structural Properties

Abstract. Phase transitions in Heusler alloys Ni(Co)-Mn-Sn were investigated with the help of the density functional theory. The lattice parameters for equilibrium magnetic states were determined using *ab initio* calculations. Possible martensitic phase transitions from the cubic $L2_1$ structure to the tetragonal $L1_0$ state were predicted. It was found martensitic phase transition does not accompany with magnetic ordering change in case of ternary alloy $Ni_2Mn_{1.5}Sn_{0.5}$. However, Co-doping leads to the structural transition accompanied by the change in the magnetic ordering from ferromagnetic to ferrimagnetic one.

Introduction

During the last two decades, the magnetocaloric effect (MCE) in magnetic-ordered compounds is intensively studying due to the ability of application of this effect in magnetic cooling technology. This technology is promising to replace a gas cooling one due to higher energy efficiency and environmental friendliness. There are a number of magnetic materials, in which MCE is observed, including the X_2YZ magnetic Heusler alloys. Recent papers concerning the influence of Co- and Cr-doping on magnetocaloric properties of Ni-Mn-Z (Z = Ga, In, Sn, Sb) alloys stimulate the interest in MCE problem optimization. Nowadays, information about the Co-doping effect in Ni-Mn-Z can be found in papers [1-14]. Adding of Co atoms leads to enhancing of ferromagnetism in austenitic phase and sharp change in phase transition temperature as well as reverse MCE detected in $Ni_{45.2}Co_{5.1}Mn_{36.7}In_{13}$ compound [1]. However, the problem with thermal hysteresis appears, which in turns leads to irreversible MCE. One of the ways of hysteresis decreasing is external pressure application. Thus, cooling capacity can be increased by simultaneously changing of the magnetic field and external pressure.

Therefore, Co-doping might influence significantly on the structural, magnetic, and magnetocaloric Heusler alloys properties. In this regard, the systematic theoretical study of the equilibrium state and possible structural transitions in $Ni_2Mn_{1.5}Sn_{0.5}$, $Ni_{1.75}Co_{0.25}Mn_{1.5}Sn_{0.5}$, and $Ni_{1.625}Co_{0.375}Mn_{1.625}Sn_{0.375}$ Heusler alloys were carried out with the help of density functional theory. It was found that there is a big jump in the magnetic moment in the $Ni_{1.75}Co_{0.25}Mn_{1.5}Sn_{0.5}$ alloy.

Calculations Details

Three nonstoichiometric compositions were considered, they are $Ni_{16}Mn_{12}Sn_4$, $Ni_{14}Co_2Mn_{12}Sn_4$, and $Ni_{13}Co_3Mn_{13}Sn_3$. All calculations were conducted in the framework of quantum chemical

Published under license by Materials Research Forum LLC.

Shape Memory Alloys – SMA 2018   Materials Research Forum LLC
Materials Research Proceedings **9** (2018) 98-103   doi: http://dx.doi.org/10.21741/9781644900017-18

simulation with the help of Vienna *Ab initio* Simulation Package (VASP) [15,16] using density functional theory (DFT) [17,18] and projector augmented-wave method (PAW) [16,24]. The generalized gradient approximation (GGA) scheme within Perdew, Burke, and Ernzerhof (PBE) [20] was considered for the exchange-correlation potential. Structural optimization was performed for 32-atom supercells. The first Brillouin zone was divided into 8×8×8 mesh according to the Monkhorst-Pack scheme [21]. The cutoff energy for the plane-wave basis was set at 400–750 eV. Geometrical optimization was carried out until forces acting on atoms equal to 0.01 meV/Å. Next electronic configurations were used for calculations: $Ni(3d^8 3p^6 4s^2)$, $Co(3d^8 4s^1)$, $Mn(3p^6 3d^5 4s^2)$, and $Sn(4d^{10} 5p^2 5s^2)$.

The simulation was done for $L2_1$ cubic structure (*Fm-3m*, № 225) with Cu_2MnAl as a prototype. Such structure with the generic formula X_2YZ consists of four interpenetrating fcc lattices, in which two X, Y, and Z atoms locate on *8c* ((1/4, 1/4, 1/4) and (3/4, 3/4, 3/4)), *4a* (0, 0, 0), and *4b* (1/2, 1/2, 1/2) Wyckoff position, correspondingly.

In case of $Ni_{16}Mn_{12}Sn_4$ and $Ni_{14}Co_2Mn_{12}Sn_4$, Mn excess atoms (Mn_2) take place of Sn, Co excess atoms are situated in Ni positions. In case of $Ni_{13}Co_3Mn_{13}Sn_3$, 7 atoms of Mn located in Sn sites, while 2 atoms of Sn (Sn_2) are in Mn positions. In order to implement this method of additional atoms placement, the USPEX software package was used [22,23]. This package has an option for obtaining special quasirandom structure (SQS). This option allows obtaining structures with an optimal arrangement of additional atoms at the positions of "parent" atoms.

Ferromagnetic (FM) configuration was set by magnetic moments of all Mn atoms are parallel to each other. Ferrimagnetic (FIM) configuration was assigned by magnetic moments of Mn atoms at the Sn (Mn_2) sites are opposite in the direction to the magnetic moments of the Mn atoms located in their sublattice (Mn_1). Thus, the ferromagnetic spin configurations, as well as ferrimagnetic ones, can be realized in studied alloys.

Results and Discussion

As an initial configuration, $Ni_{16}Mn_{12}Sn_4$ ($Ni_2Mn_{1.5}Sn_{0.5}$) compound was chosen. The energies of pointed configurations were calculated using VASP. The dependence of the energy on the lattice parameter and on the degree of tetragonal distortion is shown in Fig. 1.

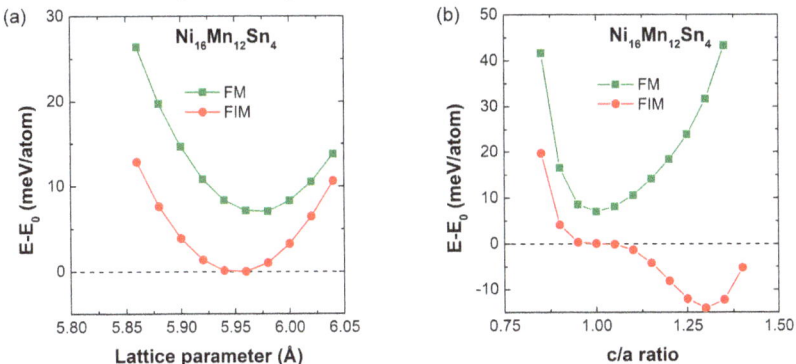

Figure 1. The dependence of the energy of $Ni_{16}Mn_{12}Sn_4$ alloy on (a) lattice parameter and (b) the degree of tetragonality for FM and FIM configurations.

It is seen from the Fig. 1 that FIM state is energetically favorable. It was found that the martensitic transition from the cubic ferrimagnetic state to the tetragonal ferrimagnetic one with $c/a = 1.3$ practically without a jump in the value of the magnetic moment ($\mu \approx 2.1\ \mu_B$) is possible in the considered alloy. The magnetic moment is 6.4 μ_B in the ferromagnetic phase. For all considered compositions, the value of the martensitic transition temperature was estimated as

Shape Memory Alloys – SMA 2018   Materials Research Forum LLC
Materials Research Proceedings 9 (2018) 98-103   doi: http://dx.doi.org/10.21741/9781644900017-18

$T_m \approx (E_a - E_m)/k_B$, where E_a and E_m are the energies of austenite and martensite, correspondingly, k_B is the Boltzmann constant. It was found, T_m is ≈ 164 K for $Ni_{16}Mn_{12}Sn_4$. Obtained results are in a good agreement with results presented in [12].

Let's consider $Ni_{14}Co_2Mn_{12}Sn_4$ ($Ni_{1.75}Co_{0.25}Mn_{1.5}Sn_{0.5}$) alloy and study the influence of the Ni atoms substitution by Co on the martensitic transformations.

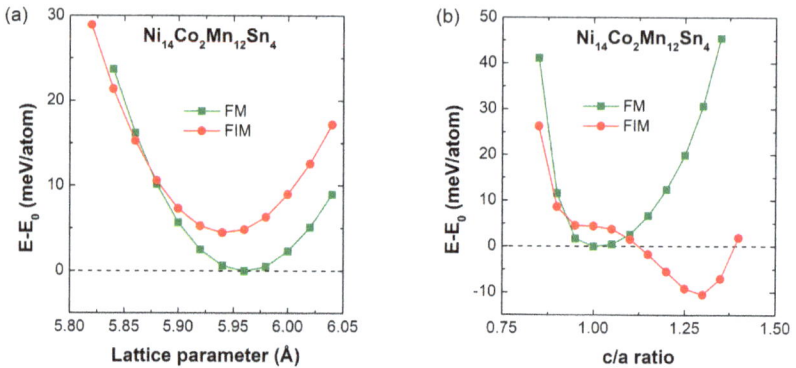

Figure 2. The dependence of the energy of the $Ni_{14}Co_2Mn_{12}Sn_4$ alloy on (a) lattice parameter and (b) the degree of tetragonality for FM and FIM configurations.

The variation of the total energy as a function of lattice parameter and degree of tetragonality is presented in Fig. 2. It is seen from this Fig. 2 (b) that FM state is more energetically favorable in cubic phase ($c/a = 1$), while tetragonal FIM state becomes favorable at $c/a > 1$. The energy minimum exists at $c/a = 1.3$. Consequently, the martensitic phase transition from FM austenitic phase to FIM martensitic one with $c/a = 1.3$ is possible in the considered alloy. The magnetic moment per f.u. of austenite is $\mu \approx 6.5\ \mu_B$, while the magnetic moment of martensite is $\mu \approx 2.4\ \mu_B$. The evaluation of the martensitic phase transition temperature which is determined from the difference between the energies of austenite and martensite (in eV/atom) gives $T_m \approx 121$ K. Therefore, the transition from FM austenite to FIM martensite with a rather big jump of the magnetic moment is possible in $Ni_{14}Co_2Mn_{12}Sn_4$ Heusler alloy.

The studying of the $Ni_{13}Co_3Mn_{13}Sn_3$ ($Ni_{1.625}Co_{0.375}Mn_{1.625}Sn_{0.375}$) is especially interesting due to the fact that the magnetocaloric effect with the big cooling effect was found experimentally [13] in similar in composition $Ni_{12.8}Co_{3.2}Mn_{12.8}Sn_{3.2}$ alloy. It is seen from the dependence of the energy of the $Ni_{13}Co_3Mn_{13}Sn_3$ alloy on the degree of tetragonality presented in Fig. 3(b) that FM state is energetically favorable in case of cubic phase. The energy minimum of the tetragonal phase appears at $c/a = 1.25$, which indicates the martensitic phase transition from FM austenitic to FIM martensitic phase with this ratio of the tetragonal distortion. The magnetic moment per f.u. of austenite is $\mu \approx 7.21\ \mu_B$, while the magnetic moment of martensite is $\mu \approx 0.53\ \mu_B$. The magnetic moment jump at martensitic transition is $\mu \approx 6.68\ \mu_B$, which is the biggest value among all considered alloys. The evaluation showed that the value of the martensitic phase transition temperature is ≈ 30 K.

Materials Research Forum LLC
doi: http://dx.doi.org/10.21741/9781644900017-18

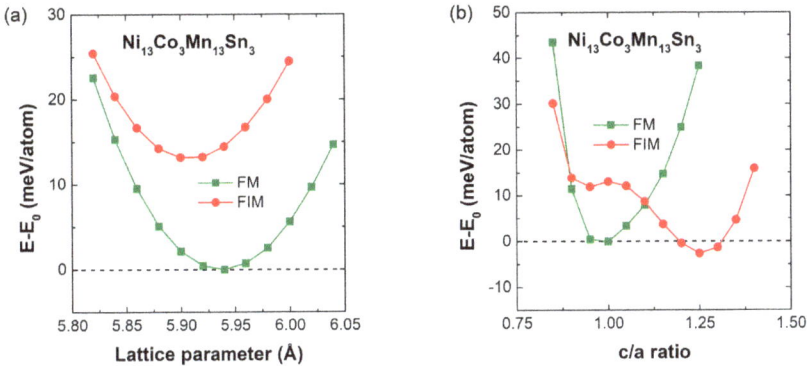

Figure 3. The dependence of the energy of the $Ni_{13}Co_3Mn_{13}Sn_3$ alloy on (a) lattice parameter and (b) the degree of tetragonality for FM and FIM configurations.

Conclusion

In this work, the research of phase transitions in ternary and quaternary Ni(Co)-Mn-Sn Heusler alloys using density functional theory was conducted. It was found structural transition is possible for $Ni_2Mn_{1.5}Sn_{0.5}$ Heusler compound, while the transition from ferromagnetic austenite to ferrimagnetic martensite is possible in considered Co-doping $Ni_{14}Co_2Mn_{12}Sn_4$ and $Ni_{13}Co_3Mn_{13}Sn_3$ alloys. Moreover, it does not accompany with the jump of the magnetic moment in case of $Ni_{16}Mn_{12}Sn_4$, while for $Ni_{14}Co_2Mn_{12}Sn_4$ this jump is $\mu \approx 4.2\ \mu_B$ and for $Ni_{13}Co_3Mn_{13}Sn_3$ – $\mu \approx 6.68\ \mu_B$. Therefore, the most suitable for the magnetic cooling is $Ni_{13}Co_3Mn_{13}Sn_3$ compound.

Acknowledgements

This work was supported by Russian Science Foundation No. 17-72-20022.

References

[1] J. Liu, T. Gottschall, K.P. Skokov, J.D. Moore, O. Gutfleisch, Giant magnetocaloric effect driven by structural transitions, Nature Mater. 11 (2012) 620-626. https://doi.org/10.1038/nmat3334

[2] T. Gottschall, K.P. Skokov, B. Frincu, O. Gutfleisch, Large reversible magnetocaloric effect in Ni-Mn-In-Co, Appl. Phys. Lett. 106 (2015) 021901. https://doi.org/10.1063/1.4905371

[3] D.Y. Cong, S. Roth, L. Schultz, Magnetic properties and structural transformations in Ni–Co–Mn–Sn multifunctional alloys, Acta Materialia 60 (2012) 5335-5351. https://doi.org/10.1016/j.actamat.2012.06.034

[4] S. Fabbrici, G. Porcari, F. Cugini, Ma. Solzi, J. Kamarad, Z. Arnold, R. Cabassi, F. Albertini, Co and In doped Ni-Mn-Ga magnetic shape memory alloys: A thorough structural, magnetic and magnetocaloric study, Entropy 16 (2014) 2204-2222. https://doi.org/10.3390/e16042204

[5] F. Guillou, P. Courtois, L. Porcar, P. Plaindoux, D. Bourgault, V. Hardy, Calorimetric investigation of the magnetocaloric effect in Ni45Co5Mn37.5In12.5, J. Phys. D: Appl. Phys. 45 (2012) 255001. https://doi.org/10.1088/0022-3727/45/25/255001

[6] J.A. Monroe, I. Karaman, B. Basaran, W. Ito, R.Y. Umetsu, R. Kainuma, K.Koyama, Y.I. Chumlyakov, Direct measurement of large reversible magnetic-field-induced strain in Ni-

Co-Mn-In metamagnetic shape memory alloys, Acta Mater. 60 (2012) 6883-6891. https://doi.org/10.1016/j.actamat.2012.07.040

[7] D.Y. Cong, S. Roth, Y.D. Wang, Superparamagnetism and superspin glass behaviors in multiferroic NiMn-based magnetic shape memory alloys, Phys. Status Solidi B. 251 (2014) 2126-2134.

[8] R. Kainuma, Y. Imano, W. Ito, Y. Sutou, H. Morito, S. Okamoto, O. Kitakami, K. Oikawa, A. Fujita, T. Kanomata, K. Ishida, Magnetic-field-induced shape recovery by reverse phase transformation, Nature 439 (2006) 957-960. https://doi.org/10.1038/nature04493

[9] R. Das, S. Sarma, A. Perumal, A. Srinivasan, Effect of Co and Cu substitution on the magnetic entropy change in Ni46Mn43Sn11 alloy, J. Appl. Phys. 109 (2011) 07A901.

[10] V.K. Sharma, M.K. Chattopadhyay, L.S. Sharath Chandra and S.B. Roy, Elevating the temperature regime of the large magnetocaloric effect in a NiMnIn alloy towards room temperature, J. Phys. D: Appl. Phys. 44 (2011) 145002. https://doi.org/10.1088/0022-3727/44/14/145002

[11] V.K. Sharmaa, M.K. Chattopadhyay, L.S. Sharath Chandra, Ashish Khandelwal, R.K. Meena, S.B. Roy, Scaling of the isothermal entropy change and magnetoresistance in Ni-Mn-In based off-stoichiometric Heusler alloys, Eur. Phys. J. Appl. Phys. 62 (2013) 30601. https://doi.org/10.1051/epjap/2013120256

[12] V. Sánchez-Alarcos, V. Recarte, J.I. Pérez-Landazábal, J.R. Chapelon, J.A. Rodríguez-Velamazán, Structural and magnetic properties of Cr-doped Ni-Mn-In metamagnetic shape memory alloys, J. Phys. D: Appl. Phys. 44 (2011) 395001. https://doi.org/10.1088/0022-3727/44/39/395001

[13] M. Khan, J. Jung, S.S. Stoyko, A. Mar, A. Quetz, T. Samanta, I. Dubenko, N. Ali, S. Stadler, K.H. Chow, The role of Ni-Mn hybridization on the martensitic phase transitions in Mn-rich Heusler alloys, Appl. Phys. Lett. 100 (2012) 172403. https://doi.org/10.1063/1.4705422

[14] M. Khan, I. Dubenko, S. Stadler, J. Jung, S.S. Stoyko, A. Mar, A. Quetz, T. Samanta, N. Ali, K.H. Chow, Enhancement of ferromagnetism by Cr doping in Ni-Mn-Cr-Sb Heusler alloys, Appl. Phys. Lett. 102 (2013) 112402. https://doi.org/10.1063/1.4795627

[15] G. Kresse, J. Furthmüller, Efficient iterative schemes for ab initio total-energy calculations using a plane-wave basis set, Phys. Rev. B 54 (1996) 11169-11186. https://doi.org/10.1103/PhysRevB.54.11169

[16] G. Kresse, D. Joubert, From ultrasoft pseudopotentials to the projector augmented-wave method, Phys. Rev. B 59 (1999) 1758-1775. https://doi.org/10.1103/PhysRevB.59.1758

[17] P. Hohenberg, W. Kohn, Inhomogeneous Electron Gas, Phys. Rev. 136 (1964) B864-B871. https://doi.org/10.1103/PhysRev.136.B864

[18] W. Kohn, L.J. Sham, Self-Consistent Equations Including Exchange and Correlation Effects, Phys. Rev. 140 (1965) A1133-A1138. https://doi.org/10.1103/PhysRev.140.A1133

[19] P.E. Blöchl, Projector augmented-wave method, Phys. Rev. B 50 (1994) 17953-17979. https://doi.org/10.1103/PhysRevB.50.17953

[20] J.P. Perdew, K. Burke, and M. Enzerhof, Generalized Gradient Approximation Made Simple, Phys. Rev. Lett. 77 (1996) 3865-3868. https://doi.org/10.1103/PhysRevLett.77.3865

[21] H.J. Monkhorst, J.D. Pack, Special points for Brillouin-zone integrations, Phys. Rev. B 13 (1976) 5188-5192. https://doi.org/10.1103/PhysRevB.13.5188

Shape Memory Alloys – SMA 2018   Materials Research Forum LLC
Materials Research Proceedings **9** (2018) 98-103   doi: http://dx.doi.org/10.21741/9781644900017-18

[22] C.W. Glass, A.R. Oganov, N. Hansen, USPEX – Evolutionary crystal structure prediction, Comp. Phys. Comm. 175 (2006) 713-720. https://doi.org/10.1016/j.cpc.2006.07.020

[23] A.O. Lyakhov, A.R. Oganov, H.T. Stokes, Q. Zhu, New developments in evolutionary structure prediction algorithm USPEX, Comp. Phys. Comm. 184 (2013) 1172-1182. https://doi.org/10.1016/j.cpc.2012.12.009

Shape Memory Alloys – SMA 2018
Materials Research Proceedings 9 (2018) 104-108

Materials Research Forum LLC
doi: http://dx.doi.org/10.21741/9781644900017-19

The Influence of Exchange-Correlation Functionals on the Ground State Properties of Ni$_2$Mn(Ga,Sn) and Fe$_2$(Ni,V)(Ga,Al) Heusler Alloys

Olga N. Miroshkina[1,a*], Mikhail A. Zagrebin[1,2,b], Oksana O. Pavlukhina[1,c], Vladimir V. Sokolovskiy[1,2,d], Vasiliy D. Buchelnikov [1,2,e]

[1]Chelyabinsk State University, 129 Br. Kashirinykh St., Chelyabinsk 454001, Russia

[2]National University of Science and Technology "MISiS", 4 Leninskiy Prospect, Moscow 119991, Russia

[a]miroshkina.on@yandex.ru, [b]miczag@mail.ru, [c]pavluhinaoo@mail.ru, [d]vsokolovsky84@mail.ru, [e]buche@csu.ru

*corresponding author

Keywords: *Ab Initio* Calculations, Heusler Alloys, Exchange-Correlation Functionals

Abstract. In this paper, the comparison of generalized-gradient and meta-generalized-gradient approximations was conducted by the example of Ni$_2$Mn(Ga,Sn) and Fe$_2$(Ni,V)(Ga,Al) full Heusler alloys. The analysis of structural, electronic, and magnetic properties of these alloys showed the results obtained with two mentioned above functionals are differ from each other. The results of well-tested generalized-gradient are in a good agreement with the experimental data. Thus, the research reveals that meta-generalized-gradient approximation is needed in follow-up revision.

Introduction

Heusler alloys are of significant interest among scientific community due to the number of the distinguished properties such as shape memory effect, effects of superelasticity and superplasticity, giant magnetocaloric effect, giant magnetoresistance and magnetostrain, etc. Therefore, full Heusler alloys Ni$_2$Mn(Ga,Sn) and Fe$_2$(V,Ni)(Al,Ga) are promising materials for shape memory, spintronic application [1].

Nowadays, the Perdew-Burke-Ernzerhof (PBE) generalized-gradient approximation is one of the widely used and well-tested approximation [2]. However, the accuracy of the generalized-gradient approximations (GGA), to which PBE is refer, is still limited. To improve the accuracy, meta-GGA was proposed. In contrast to GGA, meta-GGA takes into account additional interactions. SCAN is one of functionals of such type [3]. In present work, PBE and SCAN approximations are compared by an example of structural properties of Fe$_2$-based and Ni$_2$-based full Heusler structures.

Calculations Details

In this work, two Ni-based (Ni$_2$MnGa and Ni$_2$MnSn) and two Fe-based (Fe$_2$VAl and Fe$_2$NiGa) stoichiometric compositions were studied by means of *ab initio* methods. Calculations were conducted for 4-atom cells with L2$_1$ cubic structure (space symmetry group *Fm-3m*, No. 225). All compounds have ferromagnetic configurations, where magnetic moments of all atoms are positive.

Density functional theory (DFT) [4-6] implemented in the Vienna *Ab initio* Simulation Package (VASP) [7, 8] was used for the structural ground state calculations. Projector augmented wave (PAW) method was applied to describe interaction between ions and electrons. The kinetic energy cut-off was 550 eV, while the kinetic energy cut-off for the augmentation charges was 800 eV. Brillouin zone sampling was carried out according to the Monkhorst-Pack

Published under license by Materials Research Forum LLC.

Shape Memory Alloys – SMA 2018   Materials Research Forum LLC
Materials Research Proceedings **9** (2018) 104-108   doi: http://dx.doi.org/10.21741/9781644900017-19

grid [9]. The *k*-sampling was performed with 8×8×8 *k*-points. Using the results of electronic relaxation, equilibrium lattice parameters of studying alloys were estimated in accordance with the Birch–Murnagahan equation of state.

Both the GGA potential in PBE approximation and meta-GGA potential in SCAN approximation, which in contrast to PBE takes into account additional interactions, were used in our calculations.

Results and Discussion

In order to calculate equilibrium lattice parameter a_0, the difference between total and equilibrium energies of the studying structure was calculated.

It is seen from Fig. 1, that equilibrium lattice parameter obtained from the SCAN calculations is smaller than a_0 from PBE. In case of PBE calculations, the energy minimum of the tetragonal phase exists at $c/a = 1.25$. This value differs from SCAN result, which gives $c/a = 1.2$ and is in a good agreement with the experimental study presented in Ref. [10].

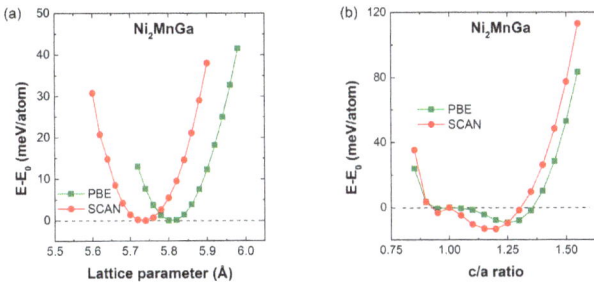

Figure 1. The dependence of the energy difference on the (a) lattice parameter and (b) degree of tetragonality (c/a ratio) for Ni₂MnGa Heusler alloy.

The results of *ab initio* calculations for Ni₂MnSn are presented in Fig. 2. The value of equilibrium lattice obtained with SCAN is lower than obtained with PBE one anв experimental value. The analysis of the tetragonal distortions possibility shows that there is no martensitic transformation in Ni₂MnSn Heusler alloy and this conclusion is common for both PBE and SCAN approximations.

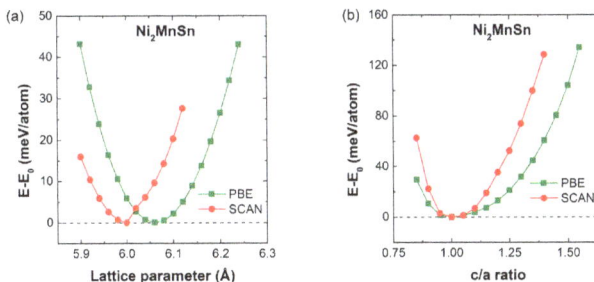

Figure 2. The dependence of the energy difference on the (a) lattice parameter and (b) degree of tetragonality (c/a ratio) for Ni₂MnSn Heusler alloy.

The results of calculations for Fe₂-based alloys are presented in Fig. 3 and Fig. 4. As surely as in case of Ni₂Mn-based compounds, the values of equilibrium lattice parameters for both Fe₂VAl and Fe₂NiGa are higher in case of PBE calculations than in case of SCAN. For Fe₂VAl, GGA PBE approximation gives the better agreement with the experimental data [11]. The results of

Shape Memory Alloys – SMA 2018   Materials Research Forum LLC
Materials Research Proceedings **9** (2018) 104-108   doi: http://dx.doi.org/10.21741/9781644900017-19

our calculations are in a good agreement with the theoretical study [12]. The behavior of the energy differences versus tetragonal distortions obtained with PBE and SCAN are qualitative agreement to each other.

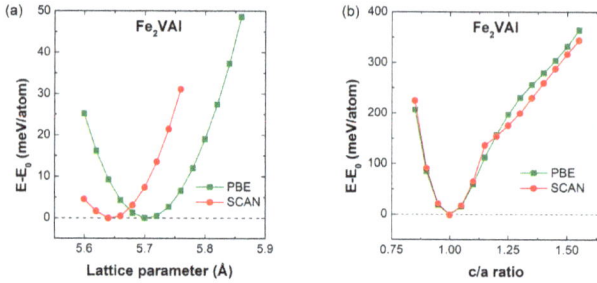

Figure 3. The dependence of the energy difference on the (a) lattice parameter and (b) degree of tetragonality (c/a ratio) for Fe_2VAl Heusler alloy.

The obtained with the help of PBE calculations equilibrium lattice parameter corresponds to the experimental value unlike SCAN result.

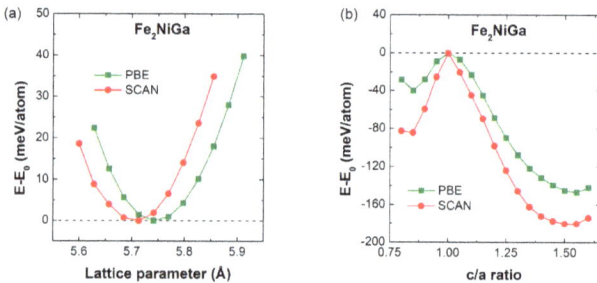

Figure 4. The dependence of the energy difference on the (a) lattice parameter and (b) degree of tetragonality (c/a ratio) for Fe_2NiGa Heusler alloy.

Calculated with the help of *ab initio* methods using PBE and SCAN approximations equilibrium lattice parameter a_0 and total magnetic moment μ for all considered alloys are summarized in Table 1. It is seen from this table that values of PBE calculations are in better agreement with experimental data, while SCAN results give underestimated values of a_0 and overestimated values of μ.

Total densities of states (DOS) for Ni_2MnGa, Ni_2MnSn, Fe_2NiGa, and Fe_2VAl are presented in Fig.5(a)-(d). It is seen from Fig. 4 that DOS curves of Ni_2MnGa calculated with SCAN are shifted symmetrically relative to Fermi level. The analysis of DOS curves showed Ni_2MnGa demonstrates metallic behavior as well as Ni_2MnSn. However, the difference between PBE and SCAN results is much greater in case of Ni_2MnSn. In contrast to Ni_2-based compounds, Fe_2VAl demonstrates semi-metallic behavior, moreover, the energy gap on DOS curves of SCAN is wider than PBE. DOS curves for spin-up and spin-down states are symmetric both for PBE and SCAN calculations, which explains the zero magnetic moment.

Table 1. The equilibrium lattice parameter obtained with PBE and SCAN for Ni_2MnGa, Ni_2MnSn, Fe_2NiGa, and Fe_2VAl in comparison with the available experimental data [11,13-15].

	Equilibrium lattice parameter [Å]			Total magnetic moment [μ_B]		
	PBE	SCAN	experiment	PBE	SCAN	experiment
Ni_2MnGa	5.81	5.74	5.83	4.08	4.72	4.17
Ni_2MnSn	6.06	5.99	6.05	4.04	6.39	4.05
Fe_2NiGa	5.74	5.70	5.78	4.86	6.03	3.21
Fe_2VAl	5.70	5.64	5.76	0.00	0.00	–

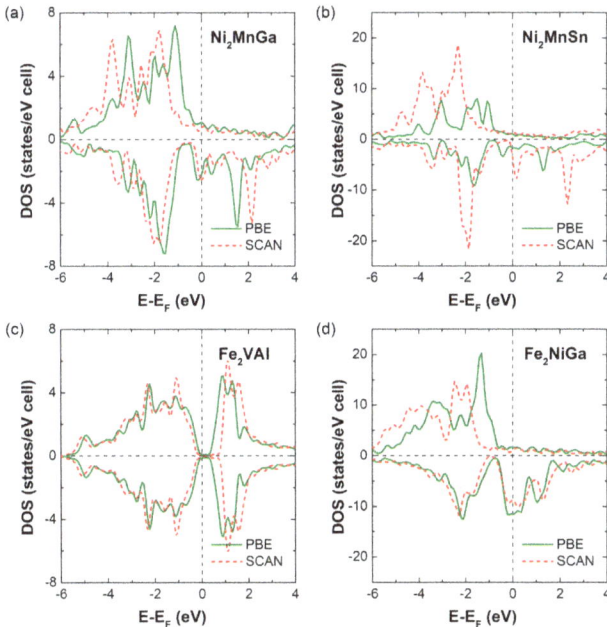

Figure 5. Total DOS curves calculated using PBE and SCAN approximations for (a) Ni_2MnGa, (b) Ni_2MnSn, (c) Fe_2VAl, and (d) Fe_2NiGa Heusler alloy.

Conclusion

In present work, test of meta-GGA SCAN functional was carried out. Verification was made in comparison with well-approved PBE approximation. Conducted with the help of *ab initio* methods research shows that application of SCAN approximation leads to underestimation in equilibrium lattice parameter calculations and overestimation of total magnetic moments. It would be reasonable to conclude that SCAN approximation is needed in some improvements.

Acknowledgements

This work was supported by Russian Foundation for Basic Research 18-32-00507 (Ni_2MnGa properties calculations) and Russian Science Foundation No. 17-72-20022 (Ni_2MnSn, Fe_2VAl,

Shape Memory Alloys – SMA 2018   Materials Research Forum LLC
Materials Research Proceedings **9** (2018) 104-108   doi: http://dx.doi.org/10.21741/9781644900017-19

and Fe$_2$NiGa properties calculations). O.M. acknowledges Young Scientist Support Fund of Chelyabinsk State University.

References

[1] C. Felser, A. Hirohata (Eds.), Heusler Alloys: Properties, Growth, Applications, Springer, New York, 2016. https://doi.org/10.1007/978-3-319-21449-8

[2] J.P. Perdew, K. Burke, M. Enzerhof, Generalized Gradient Approximation Made Simple, Phys. Rev. Lett. 77 (1996) 3865-3868. https://doi.org/10.1103/PhysRevLett.77.3865

[3] J. Sun, M. Marsman, G.I. Csonka, A. Ruzsinszky, P. Hao, Y.-S. Kim, G. Kresse, J.P. Perdew, Self-consistent meta-generalized gradient approximation within the projector-augmented-wave method, Phys. Rev. B 84 (2011) 035117. https://doi.org/10.1103/PhysRevB.84.035117

[4] W. Kohn, L.J. Sham, Self-Consistent Equations Including Exchange and Correlation Effects, Phys. Rev. 140 (1965) A1133-A1138. https://doi.org/10.1103/PhysRev.140.A1133

[5] R.G. Parr, W. Yang, Density Functional Theory of Atoms and Molecules, Oxford University Press, Oxford, 1989.

[6] J.P. Perdew, S. Kurth, A Primer in Density Functional Theory, Springer, New York, 2003.

[7] G. Kresse, J. Furthmüller, Efficient iterative schemes for ab initio total-energy calculations using a plane-wave basis set, Phys. Rev. B 54 (1996) 11169-11186. https://doi.org/10.1103/PhysRevB.54.11169

[8] G. Kresse, D. Joubert, From ultrasoft pseudopotentials to the projector augmented-wave method, Phys. Rev. B 59 (1999) 1758-1775. https://doi.org/10.1103/PhysRevB.59.1758

[9] H.J. Monkhorst, J.D. Pack, Special points for Brillouin-zone integrations, Phys. Rev. B 13 (1976) 5188-5192. https://doi.org/10.1103/PhysRevB.13.5188

[10] A. Sozinov, A.A. Likhachev, N. Lanska, K. Ullakko, Giant magnetic-field-induced strain in NiMnGa seven-layered martensitic phase, Appl. Phys. Lett. 80 (2002) 1746-1748. https://doi.org/10.1063/1.1458075

[11] P.J. Webster, K.R.A. Ziebeck, The paramagnetic properties of Heusler alloys containing iron, Phys. Lett. A 98 (1983) 51-53. https://doi.org/10.1016/0375-9601(83)90543-1

[12] S.S. Shastri, S.K. Pandey, A comparative study of different exchange-correlation functionals in understanding structural, electronic and thermoelectric properties of Fe$_2$VAl and Fe$_2$TiSn compounds, Comp. Mater. Sci. 143 (2018) 316-324. https://doi.org/10.1016/j.commatsci.2017.10.053

[13] P.J. Webster, K.R.A. Ziebeck, Alloys and Compounds of 3d-Elements with Main Group Elements, in: H.R.J. Wijn (Ed.), Magnetic Properties of Metals, Springer, Berlin, 1988, pp. 75-184.

[14] P.J. Webster, K.R.A. Ziebeck, S.L. Town, M.S. Peak, Magnetic order and phase transformation in Ni$_2$MnGa, Philos. Mag. B 49 (1984) 295-310. https://doi.org/10.1080/13642817408246515

[15] K.H.J. Buschow, P.G. van Engen, Magnetic and Magneto-Optical Properties of Heusler Alloys Based on Aluminium and Gallium, J. Magn. Magn. Mater. 25 (1981) 90-96. https://doi.org/10.1016/0304-8853(81)90151-7

Shape Memory Alloys – SMA 2018
Materials Research Proceedings 9 (2018) 109-113

Materials Research Forum LLC
doi: http://dx.doi.org/10.21741/9781644900017-20

Electronic, Structural, and Magnetic Properties of the FeRh₁₋ₓPtₓ (x = 0.875 and 1)

Oksana O. Pavlukhina[1,a*], Vasily D. Buchelnikov[1,2,b], Vladimir V. Sokolovskiy[1,2,c], Mikhail A. Zagrebin[1,2,d]

[1]Chelyabinsk State University, 129 Brat'ev Kashirinykh Str., Chelyabinsk 454001, Russia

[2]National University of Science and Technology "MISiS", 4 Leninskiy Prospect, Moscow 119049, Russia

[a]pavluhinaoo@mail.ru, [b]buche@csu.ru, [c]vsokolovsky84@mail.ru, [d]miczag@mail.ru

*corresponding author

Keywords: *Ab Initio* Study, Supercell Approach, Ferro- And Antiferromagnetic Orders, Iron-Rhodium Alloy, Density of States

Abstract. Using the *ab initio* study, we theoretically investigated the electronic, structural and magnetic properties of the FeRh₁₋ₓPtₓ ($x = 0.875$ and 1). For the Pt concentrations $x = 0.875$, the spin configuration AFM-III is stable. For FePt the ferromagnetic phase is more stable. It is shown that the equilibrium lattice parameter and the type of magnetic ordering change with increasing Pt concentration. The simulated value of the Curie temperature for compositions is close to experimental data.

Introduction

Fe-Rh alloy alloys attract increasing attention of scientists all over the world due to the possibility of their practical application in magnetic cooling, magnetic recording and spintronics devices [1-3]. In Fe-Rh alloys, a metamagnetic phase transition at temperatures close to room temperature is observed. The metamagnetic phase transition in Fe-Rh also leads to large changes in the magnetization, which causes a giant magnetocaloric effect with a change in the magnetic field.

In research works [4,5] the magnetocaloric property of Fe₄₉Rh₅₁ alloys have been measured by direct method. Phase transition is notable at the temperature close to 310 K. Adiabatic temperature change of Fe₄₉Rh₅₁ alloy was 13 K with magnetic field change up to 2 T. Magnetic order in Fe-Rh compounds is strongly dependent on elements concentration. Therefore it is important to study influence of adding third element on magnetic and structural properties of material. In work [6] studied influence of dopant Co, Pd, Ru and Pl on magnetic moment and density of states in Fe-Rh alloys. Content of the third element in FeRh₁₋ₓ(Z)ₓ alloys was small: $x = 0$–0.05. The work says about significant change of magnetic moment and density of states on Fermi level during the process of metal replacement. As regards influence of addition of Pt on structure and property of Fe-Rh, such studies are also very interesting. In works [7,8], alloys Fe-Rh-Pt were studied experimentally by using magnetometry, Mössbauer spectroscopy and X-ray diffraction. In work [7] with the help of X-ray analysis it was shown that in FeRh₁₋ₓPtₓ alloys at concentration level $x \leq 0.1$ the structure is cubic, and tetragonal phase is appeared with increasing of x.

Theoretical research helps to describe and understand the phenomena occurring in the material. Therefore, in our previous work, Fe-Rh-(Z) ($Z = $ Co, Pt) alloys were investigated using first-principles methods. The magnetic and structural properties of these materials are investigated. The concentration of the third element was 0, 0.125, 0.25 and 0.375 [9]. In [10], the structural and magnetic properties of Fe-Rh alloys with the addition of Ni and Pd were studied using first-principle methods. It is shown that the addition of the third element stimulates the

Published under license by Materials Research Forum LLC.

Shape Memory Alloys – SMA 2018   Materials Research Forum LLC
Materials Research Proceedings 9 (2018) 109-113   doi: http://dx.doi.org/10.21741/9781644900017-20

martensitic transformation. This work is devoted to theoretical studies of the structure and magnetic properties of Fe-Rh-Pt alloys by first-principles methods.

Computational details

In this work, the properties of $FeRh_{1-x}Pt_x$ alloys were studied using the Vienna *Ab Initio* Simulation Package (VASP) [11]. Calculations were carried out within the generalized gradients approximation (GGA) in the Perdew, Burke, and Ernzerhof (PBE) formulation. The plane-wave cutoff energy (E_{cutoff}) was set to 400 eV. All the studied structures had a 12×12×12 Monkhorst–Pack grid of k-points in the first Brillouin zone. The PAW potential was used. Calculations were conducted for a supercell containing 16 atoms with different initial spin configurations.

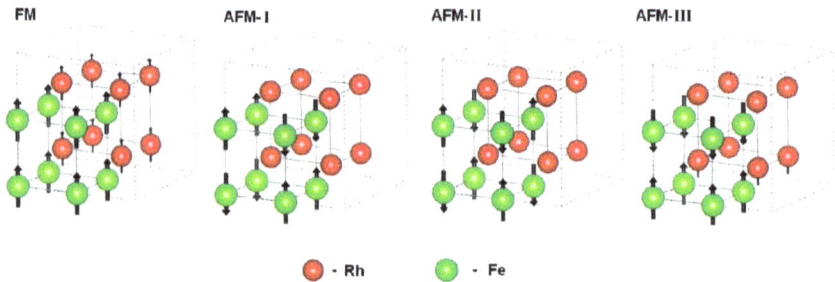

Figure 1. Magnetic spin configurations.

Energy calculations were performed for a 16-atom supercell. Rhodium and iron atoms were located at the following sites: (0; 0; 0); (1/2; 0; 0); (1/2; 1/2; 1/2) and (1/4; 1/4; 1/4); (3/4; 3/4; 3/4), respectively. One ferromagnetic (FM) state and three antiferromagnetic (AFM-I, AFM-II, and AFM-III) states were studied (see the configurations in Fig. 1). In the case of FM ordering, rhodium atoms in the considered cell had a small magnetic moment (<1 µB); if the ordering was AFM, the magnetic moment of Rh was zero. Using the optimized parameters obtained from VASP, we further calculated the exchange coupling constants J_{ij} by using the SPR-KKR package (spin-polarized-relativistic Korringa-Kohn-Rostoker). To perform this study, the coherent potential approximation (CPA) has been used. For the optimized lattice parameter, the self-consistent potential (SCF) is calculated. All calculations converged to 0.01 mRy of total energy. The maximum number of SCF iterations was taken to 200.

Results of calculations

In our previous work, Fe-Rh-based alloys were studied [10]. In this paper, it is reported that the calculated optimized lattice parameter for FeRh alloy is in a good agreement with experimental and theoretical values. The total energy for tetragonal distortion of the cubic structure along axis z was calculated. The cell volume was fixed in these calculations: $V_0 = a_0^3 \approx a^2c$. Fig. 2 illustrates energy dependence on tetragonal distortion c/a for $FeRh_{1-x}Pt_x$ system with different spin configurations. Zero ΔE values correspond to the austenitic phase in each studied compound. From data that was received it is seen that the spin configuration AFM-III is stable for $FeRh_{1-x}Pt_x$ where $x = 0.875$. For FePt the ferromagnetic phase is more stable.

Shape Memory Alloys – SMA 2018   Materials Research Forum LLC
Materials Research Proceedings **9** (2018) 109-113   doi: http://dx.doi.org/10.21741/9781644900017-20

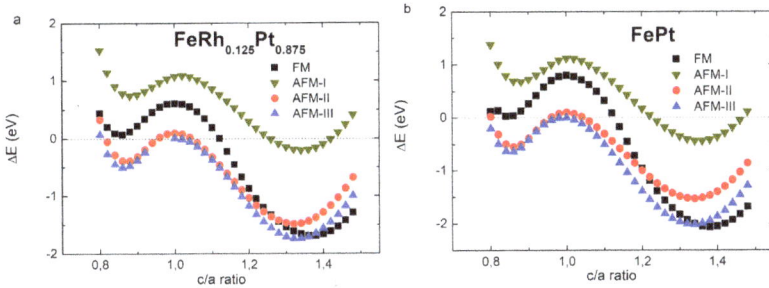

Figure 2. Dependences of total energy on tetragonal distortion c/a for FeRh$_{1-x}$Pt$_x$
(x = 0.875, 1) systems with spin configurations FM, AFM-I, AFM-II and AFM-III.

In the Table 1 results of theoretical calculations of lattice parameters for different magnetic configurations are presented. It follows from these data that the introduction of Pt atoms leads to an increase in the lattice parameter. This may be attributed to the fact that the atomic radius of Pt is larger than that of rhodium.

Table 1. Lattice parameters for different magnetic configurations

	a_0, [Å]	
	$x = 0.875$	$x = 1$
AFM-I	3.0491	3.0519
AFM-II	3.0482	3.0523
AFM- III	3.0491	3.0522
FM	3.0512	3.0595

In the next step of our study, we evaluated the Heisenberg exchange coupling constants *Jij* for FeRh$_{1-x}$Pt$_x$ (x = 0.875, 1) alloys. In Fig. 3 we present the calculated exchange coupling constants J_{ij} for tetragonal FePt alloy at its equilibrium lattice parameter.

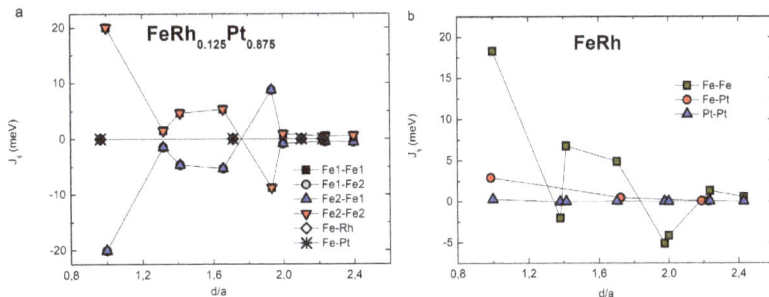

Figure 3. Exchange constants for FeRh$_{1-x}$Pt$_x$ (x = 0.875, 1) as a function of distance
in units of the lattice constant a.

The effective exchange parameters are necessary to calculate the Curie temperature using the well-established Heisenberg model in the framework of mean field approximation. The Curie temperature can be obtained by solving the system of coupled equations:

$$\left\langle s^{\mu} \right\rangle = \frac{2}{3k_B T} \sum J_0^{\mu\upsilon} \left\langle s^{\upsilon} \right\rangle. \tag{1}$$

Where $J_0^{\mu\upsilon}$ is the effective exchange parameter; μ, υ are the different sublattices; $<s^{\upsilon}>$ is the average z component of an atom spin located in the υ sublattice. The equation has non-trivial solutions if the determinant is zero. In this case, the largest eigenvalue gives the Curie temperature. Curie temperature for the FePt alloy was 785 K, for the $FeRh_{0.125}Pt_{0.875}$ alloy was 670 K. The simulation value of the Curie temperature for compositions is close to experimental data [12].

In this paper, the total and partial densities of states for the FePt and $FeRh_{0.125}Pt_{0.875}$ alloys were investigated. Fig. 4. illustrates the total density of electronic states for the FePt and $FeRh_{0.125}Pt_{0.875}$ alloys and the projected density of states for t_{2g} and e_g orbitals for FePt. The spin polarization was calculated by the formula: $P = (N\uparrow(E_F) - N\downarrow(E_F))/(N\uparrow(E_F) + N\downarrow(E_F))$, where $N\uparrow(E_F)$ and $N\downarrow(E_F)$ are the electron density states at the Fermi level (E_F) for the spin direction up and down, respectively.

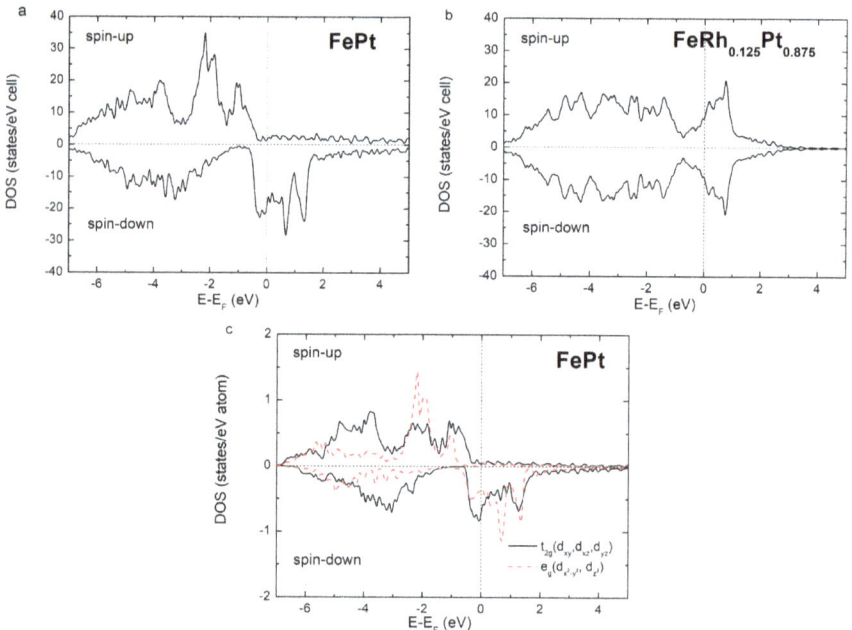

Figure 4. The total density of electronic states for the (a) FePt and (b) $FeRh_{0.125}Pt_{0.875}$ alloys and (c) the projected density of states for t_{2g} and e_g orbitals for FePt. Zero energy indicates the position of the Fermi level.

The spin polarization for the FePt alloy was approximately 80 %. It was found that the main contribution to the density of states near the Fermi level is made by the t_{2g} electrons, which give the main contribution to the spin polarization for these compounds.

Summary

In summary, we theoretically investigated the electronic, structural and magnetic properties of the $FeRh_{1-x}Pt_x$ ($x = 0.875$ and 1) using the *ab initio* calculations. The cubic phase becomes

Shape Memory Alloys – SMA 2018 Materials Research Forum LLC
Materials Research Proceedings **9** (2018) 109-113 doi: http://dx.doi.org/10.21741/9781644900017-20

unstable in these conditions. For the platinum concentrations $x = 0.875$ in martensite the AFM-III phase is more stable. With the increase of platinum content at concentrations $x = 1$ FM phase becomes more stable in martensite.

Acknowledgements

This study was supported by the Russian Science Foundation, project No. 17-72-20022-17.

References

[1] S. Cumpson, P. Hidding, R. Coehoorn, A hybrid recording method using thermally assisted writing and flux sensitive detection, IEEE Trans. Magn. 36 (2000) 2271-2275. https://doi.org/10.1109/20.908391

[2] J. Thiele, S. Maat, E. Fullerton, FeRh/FePt exchange spring films for thermally assisted magnetic recording media, Appl. Phys. Lett. 82 (2003) 2859-2861. https://doi.org/10.1063/1.1571232

[3] A. Gray, D. Cooke, P. Kruger, Electronic Structure Changes across the Metamagnetic Transition in FeRh via Hard X-Ray Photoemission, Phys. Rev. Lett. 108 (2012) 257208. https://doi.org/10.1103/PhysRevLett.108.257208

[4] M. Annaorazov, K. Asatryan, G. Myalikgulyev, S. Nikitin, A. Tishin, A. Tyurin, Alloys of the Fe-Rh system as a new class of working material for magnetic refrigerators, Cryog. 32 (1992) 867-872. https://doi.org/10.1016/0011-2275(92)90352-B

[5] A. Chirkova, K. Skokov, L. Schultz, N. Baranov, O. Gutfleisch, T. Woodcock, Giant adiabatic temperature change in FeRh alloys evidenced by direct measurements under cyclic conditions. Acta Mater. 106 (2016) 15-21. https://doi.org/10.1016/j.actamat.2015.11.054

[6] A. Jezierski, G. Borstel, Electronic and magnetic properties of Fe-Rh-TM alloys, J. Magn. Magn. Mater. 144 (2005) 81-82.

[7] S. Yuasa, H. Miyajima, Magnetic properties and phase transition in bct $FeRh_{1-x}Pt_x$, alloys, Nucl. Instrum. Methods Phys. Res. Sect. B 76 (1993) 71-73. https://doi.org/10.1016/0168-583X(93)95136-S

[8] K. Takizawa, T. Ono, H, Miyajima, Magnetic phase transitions for body-centered tetragonal $FeRh_{1-x}Pt_x$ system. J. Magn. Magn. Mater. 226–230 (2001) 572-573. https://doi.org/10.1016/S0304-8853(00)01296-8

[9] O. Pavlukhina, V. Sokolovskiy, M, Zagrebin, V. Buchelnikov, Investigation of structural and magnetic properties of Fe-Rh-(Z) (Z = Co, Pt) alloys by first principles method, V. EPJ Web Conf. 2018. 185 (2018) 05005-4. https://doi.org/10.1051/epjconf/201818505005

[10] O. Pavlukhina, V. Sokolovskiy, V. Buchelnikov, First principles study of the structural and magnetic properties of Fe(Rh, Pd) and Fe(Rh, Ni) alloys, Mater. Today: Proc. 4 (2017) 4642-4646. https://doi.org/10.1016/j.matpr.2017.04.044

[11] G. Kresse, J. Furthmuller, Efficient iterative schemes for ab initio total-energy calculations using a plane-wave basis set. Phys. Rev. B. 54 (1996) 11169. https://doi.org/10.1103/PhysRevB.54.11169

[12] S. Yuasa, H. Miyajima, Y. Otani, Magneto-Volume and Tetragonal Elongation Effects on Magnetic Phase Transitions of Body-Centered Tetragonal $FeRh_{1-x}Pt_x$, J. Phys. Soc. Jpn. 63 (1994) 3129-3144. https://doi.org/10.1143/JPSJ.63.3129

Shape Memory Alloys – SMA 2018
Materials Research Proceedings 9 (2018) 114-117

Materials Research Forum LLC
doi: http://dx.doi.org/10.21741/9781644900017-21

Ab Initio Study of Structural and Magnetic Properties of the Fe$_{0.5}$Mn$_{0.5}$Rh and Fe$_{0.375}$Mn$_{0.625}$Rh Alloys

Oksana O. Pavlukhina[1,a*], Vasily D. Buchelnikov[1,2,b], Vladimir V. Sokolovskiy[1,2,c], Mikhail A. Zagrebin[1,2,d]

[1]Chelyabinsk State University, 129 Brat'ev Kashirinykh Str., Chelyabinsk 454001, Russia

[1]National University of Science and Technology "MiSiS", 4 Leninskiy Prospect, Moscow 119049, Russia

[a]pavluhinaoo@mail.ru, [b]buche@csu.ru, [c]vsokolovsky84@mail.ru, [d]miczag@mail.ru

*corresponding author

Keywords: Iron-Rhodium Alloy, *Ab Initio* Study, Supercell Approach, Ferro and Antiferromagnetic Orders

Abstract. In the present work, the structural and magnetic properties of Fe$_{0.5}$Mn$_{0.5}$Rh and Fe$_{0.375}$Mn$_{0.625}$Rh alloys are investigated by using the density functional theory calculations within a supercell approach as implemented in the VASP code. It is shown, that the equilibrium lattice parameter change with increasing Mn concentration. The antiferromagnetic spin configuration in cubic phase is stable for Fe$_{0.5}$Mn$_{0.5}$Rh and the ferromagnetic spin configuration in cubic phase is stable for Fe$_{0.375}$Mn$_{0.625}$Rh alloys.

Introduction

Works on magnetic cooling are carried out in many laboratories in the world. An important task in magnetic cooling technology is the task of finding efficient materials. Alloys based on Fe-Rh attract more and more attention of scientists all over the world due to the possibility of their practical application in magnetic cooling, magnetic recording and spintronics devices [1,2].

The alloy Fe$_{0.5}$Rh$_{0.5}$ has a crystal structure of CsCl type: each atom Rh is located in the cube center, atoms Fe are located on each eight tops of the cube. In Fe-Rh alloys, a metamagnetic phase transition is observed at temperatures close to room temperatures. The metamagnetic phase transition in Fe-Rh also leads to large changes in the magnetization, which causes a giant magnetocaloric effect with a change in the magnetic field [3]. Magnetic order in Fe-Rh compounds is strongly dependent on elements concentration. Therefore it is important to study the influence of adding the third element on the magnetic and structural properties of the material. In work [4] the authors studied the influence of dopant Co, Pd, Ru and Pl on the magnetic moment and density of states in Fe-Rh alloys. Content of the third element in FeRh$_{1-x}$(Z)$_x$ alloys was small: $x = 0$–0.05. The work says about significant change in the magnetic moment and density of states on Fermi level during the process of metal replacement.

It is important to study the effect of adding a third element on the magnetic and structural properties of the material. Therefore, in our previous work, alloys FeRh$_{1-x}$(Z)$_x$ (Z = Mn, Ni, Pd, Co, Pt) were investigated using first-principle methods [5,6]. Here we replaced the rhodium atom. In the present work, we theoretically investigated the structural and magnetic properties of the Fe$_{0.5}$Mn$_{0.5}$Rh and Fe$_{0.375}$Mn$_{0.625}$Rh alloys.

Computational details

We theoretically investigated the structural and magnetic properties of Fe$_{0.5}$Mn$_{0.5}$Rh and Fe$_{0.375}$Mn$_{0.625}$Rh by using the *ab initio* calculations (Vienna *Ab Initio* Simulation Package) [7]. Calculations were carried out within the generalized gradients approximation in the Perdew, Burke, and Ernzerhof formulation.

Published under license by Materials Research Forum LLC.

Shape Memory Alloys – SMA 2018   Materials Research Forum LLC
Materials Research Proceedings **9** (2018) 114-117   doi: http://dx.doi.org/10.21741/9781644900017-21

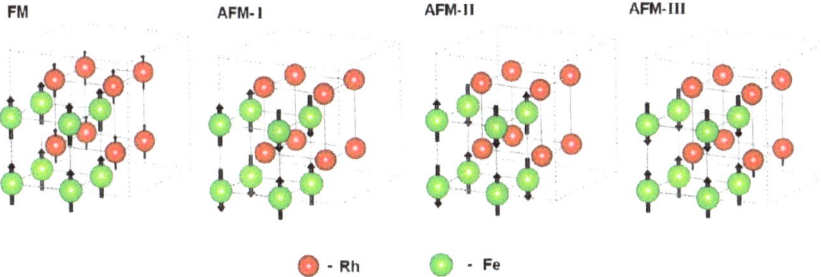

Figure 1. Different spin configurations taken into account in ab initio calculations (ferromagnetic (FM), and three kinds of antiferromagnetic states (AFM-I, AFM-II, AFM-III)).

The plane-wave cutoff energy (E_{cutoff}) was set to 400 eV. The PAW potential was used. All the studied structures had a $12 \times 12 \times 12$ Monkhorst–Pack grid of k-points in the first Brillouin zone. Calculations were conducted for a supercell containing 16 atoms with different initial spin configurations. Energy calculations were performed for a 16-atom supercell. Rhodium and iron atoms were located at the following sites: (0; 0; 0); (1/2; 0; 0); (1/2; 1/2; 1/2) and (1/4; 1/4; 1/4); (3/4; 3/4; 3/4), respectively. One ferromagnetic (FM) state and three antiferromagnetic (AFM-I, AFM-II, and AFM-III) states were studied (see the configurations in Fig. 1). In the case of FM ordering, rhodium atoms in the considered cell had a small magnetic moment (<1 μB); if the ordering was AFM, the magnetic moment of Rh was zero. PAW potentials with Fe ($3d^7$, $4s^1$), Rh ($4d^8$, $5s^1$), Mn ($3d^5$, $4s^2$) treated as valence electrons were used.

Results of calculations

Consider the effect of adding a third element on the lattice parameter. In the first step, we calculated the variations of the total energy of the 16-atom supercells $Fe_{1-x}Mn_xRh$ for different spin configurations as functions of the lattice parameter.

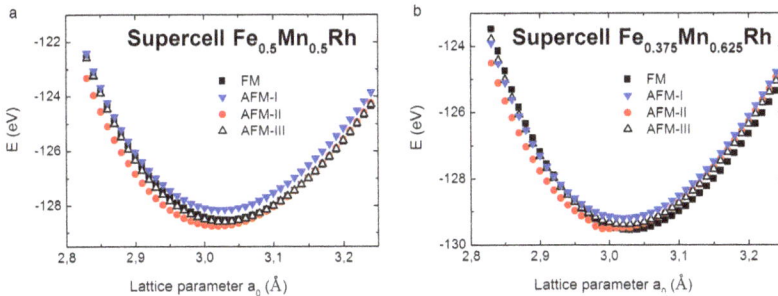

Figure 2. Energy dependence for $Fe_{0.5}Mn_{0.5}Rh$ and $Fe_{0.375}Mn_{0.625}Rh$ systems from lattice parameter for different spin configurations.

In our previous work, Fe-Rh-based alloys were studied. In this paper [5], it is reported that the calculated optimized lattice parameter for FeRh alloy is in a good agreement with experimental and theoretical values. The results of energy calculations as a function of lattice parameter for $Fe_{1-x}Mn_xRh$ systems with different spin configurations are shown in Fig. 2. From calculated data, it is seen that the spin configuration AFM-II in cubic phase is stable for $Fe_{0.5}Mn_{0.5}Rh$ and the

Shape Memory Alloys – SMA 2018   Materials Research Forum LLC
Materials Research Proceedings **9** (2018) 114-117   doi: http://dx.doi.org/10.21741/9781644900017-21

spin configuration FM in cubic phase is stable for $Fe_{0.375}Mn_{0.625}Rh$. The table 1 presents the results of theoretical calculations of the lattice parameters for different magnetic configurations.

Table 1. The results of theoretical calculations of the lattice parameters
for different magnetic configurations.

	FM	AFM-I	AFM-II	AFM-III
$Fe_{0.5}Mn_{0.5}Rh$	3.030	3.011	3.009	3.019
$Fe_{0.375}Mn_{0.625}Rh$	3.030	3.019	2.996	3.021

The total energy for tetragonal distortion of the cubic structure along axis z was calculated. The cell volume was fixed in these calculations: $V_0 = a_0^3 \approx a^2c$. Fig. 3 illustrates the dependence of energy on tetragonal distortion of c/a for the $Fe_{0.5}Mn_{0.5}Rh$ and $Fe_{0.375}Mn_{0.625}Rh$ system with different spin configurations. The obtained partial and total magnetic moments for $Fe_{1-x}Mn_xRh$ alloy with different magnetic ordering are summarized in Table 2.

Table 2. The calculated partial and total magnetic moments for $Fe_{0.5}Mn_{0.5}Rh$
and $Fe_{0.375}Mn_{0.625}Rh$ alloy with different magnetic ordering.

			Magn. mom. tot,$(\mu_B$/f.u.)	Magn. mom. Fe, (μ_B)	Magn. mom. Mn, (μ_B)	Magn. mom. Rh, (μ_B)
$x = 0.5$	$c/a =1$	FM	4.356	3.240	3.669	0.901
		AFM-I	~ 0	−3.145	3.538	−0.308
		AFM-II	~ 0	−3.124	3.469	−0.347
		AFM-III	~ 0	−3.183	3.586	−0.305
	$c/a >1$	AFM-I	~ 0	−2.927	3.344	−0.244
		AFM-II	~ 0	−2.954	3.124	−0.327
		AFM-III	~ 0	−2.915	3.400	−0.179
$x = 0.625$	$c/a =1$	FM	4.257	3.225	3.592	0.807
		AFM-I	~ 0	−3.178	±3.572	−0,234
		AFM-II	~ 0	−3.413	3.212	−0.126
		AFM-III	~ 0	−3.180	±3.578	−0.258
	$c/a >1$	AFM-I	~ 0	−2.946	±3.406	−0.182
		AFM-II	~ 0	−2.866	±3.034	−0.227

It is seen from Fig. 3(a) that the martensitic transition from the FM cubic state to AFM-II tetragonal one can exist. The curves for $Fe_{1-x}Mn_xRh$ have more noticeable minima at $c/a > 1$ with increasing Mn content. This is equivalent to an increase in the temperature of structural transformations. The martensitic phase transformation in these alloys is observed. The value of the tetragonal ratio for $Fe_{0.5}Mn_{0.5}Rh$ was $c/a = 1.26$ and for $Fe_{0.375}Mn_{0.625}Rh$ was $c/a = 1.28$.

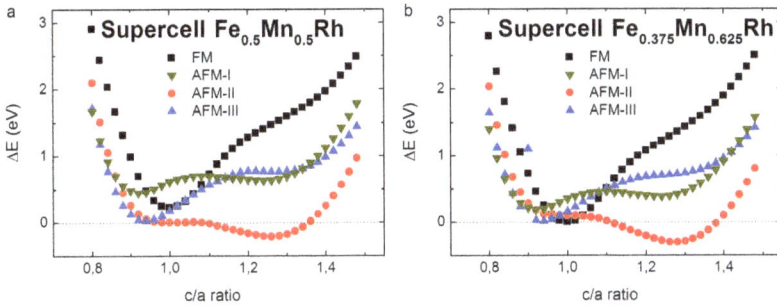

Figure 3. The dependence of energy on tetragonal distortion of c/a for the $Fe_{0.5}Mn_{0.5}Rh$ and $Fe_{0.375}Mn_{0.625}Rh$ system with different spin configurations.

Summary

In conclusion, the structural and magnetic properties of Mn-doped Fe-Rh alloys are investigated. The 16-atom supercell with different initial spin configurations including ferromagnetic and three types of antiferromagnetic orders are considered. We theoretically investigated the dependence of energy on tetragonal distortion of c/a for the $Fe_{0.5}Mn_{0.5}Rh$ and $Fe_{0.375}Mn_{0.625}Rh$ systems. The curves for $Fe_{1-x}Mn_xRh$ have more noticeable minima at $c/a > 1$ with increasing Mn content. This is equivalent to an increase in the temperature of structural transformations.

Acknowledgements

This study was supported by the Russian Science Foundation, project no. 17-72-20022.

References

[1] J. Thiele, S. Maat, E. Fullerton, FeRh/FePt exchange spring films for thermally assisted magnetic recording media, Appl. Phys. Lett. 82 (2003) 2859-2861. https://doi.org/10.1063/1.1571232

[2] A. Gray, D. Cooke, P. Kruger, Electronic Structure Changes across the Metamagnetic Transition in FeRh via Hard X-Ray Photoemission, Phys. Rev. Lett. 108 (2012) 257208. https://doi.org/10.1103/PhysRevLett.108.257208

[3] M. Annaorazov, K. Asatryan, G. Myalikgulyev, S. Nikitin, A. Tishin, A. Tyurin, Alloys of the Fe-Rh system as a new class of working material for magnetic refrigerators, Cryog. 32 (1992) 867-872. https://doi.org/10.1016/0011-2275(92)90352-B

[4] A. Jezierski, G. Borstel, Electronic and magnetic properties of Fe-Rh-TM alloys, J. Magn. Magn. Mater. 144 (2005) 81-82.

[5] O. Pavlukhina, V. Sokolovskiy, V. Buchelnikov, First principles study of the structural and magnetic properties of Fe(Rh, Pd) and Fe(Rh, Ni) alloys, Mater. Today: Proc. 4 (2017) 4642-4646. https://doi.org/10.1016/j.matpr.2017.04.044

[6] O. Pavlukhina, V. Sokolovskiy, M. Zagrebin, V. Buchelnikov, First-Principles Study of the Structure and Magnetic Properties of $Fe_8Rh_{8-x}Z_x(Z = Mn, Pt, Co; x = 1, 2, 3)$ Alloys, Phys. Solid State 60 (2018) 1134-1138. https://doi.org/10.1134/S1063783418060288

[7] G. Kresse, J. Furthmuller, Efficient iterative schemes for ab initio total-energy calculations using a plane-wave basis set, Phys. Rev. B. 54 (1996) 11169. https://doi.org/10.1103/PhysRevB.54.11169

Shape Memory Alloys – SMA 2018   Materials Research Forum LLC
Materials Research Proceedings 9 (2018) 118-121   doi: http://dx.doi.org/10.21741/9781644900017-22

First-Principles Study of Mn-rich Ni-Mn-Ga Alloys: Effect of Disorder on Martensitic Transformation

Yulia A. Sokolovskaya[1,a], Vladimir V. Sokolovskiy[1,b*], Mikhail A. Zagrebin[1,c],
Vasiliy D. Buchelnikov[1,d]

[1]Chelyabinsk State University, 129 Brat'ev Kashirinykh Str., Chelyabinsk 454001, Russia

[a]sya2890@mail.ru, [b]vsokolovsky84@mail.ru, [c]miczag@mail.ru, [c]buche@csu.ru

*corresponding author

Keywords: First-Principles Calculations, Heusler Alloys, Martensitic Transformation, Structural and Chemical Disorder

Abstract. In this work, we study theoretically the effect of chemical and structural disorder on martensitic transformation in Mn-rich Ni-Mn-Ga alloys by first-principles calculations. The both chemical and structural disorders in 16-atom supercell are assumed. For $Ni_2Mn_{1+x}Ga_{1-x}$ alloys with $0 < x < 1.5$, the staggered crystal structure of martensite with ferromagnetic order is found to be more energetically stable while with a further increase in Mn content the layered crystal structure becomes favorable in martensite phase with ferromagnetic order.

Introduction

Nowadays, the class of ferromagnetic shape memory Ni-Mn-Ga Heusler alloys attracts a great attention of scientific society due to intriguing properties. For instance, these systems display a significant shape memory [1], magnetostriction [2] and super elasticity [3] properties as well as magnetocaloric effect [4]. The most of unique features are caused by the connection between atomic and magnetic order and can be of interest for both experimental and theoretical investigations.

The high-temperature crystal structure of prototype Ni_2MnGa consists of four interpenetrated face-centered cubic sublattices of each element, which transforms martensitically to the low-temperature tetragonal structure through a series of modulated structures upon cooling below $T_m \approx 200$ K. On the other hand, the ferro-paramagnetic transition realizes in cubic austenite phase at $T_C \approx 376$ K [5]. According to the recent phase diagram of rapidly-quenched Mn-rich $Ni_2Mn_{1+x}Ga_{1-x}$ alloys, with the deviation from stoichiometry the coexistence of cubic $L2_1$ austenite, modulated 5M and 7M phases and tetragonal $L1_0$ martensite has been reported by Çakir et al. [6]. They have shown that depending on composition, the martensitic transformation occurs in the following sequences $L2_1 \rightarrow 7M \rightarrow L1_0$, $L2_1 \rightarrow 5M \rightarrow 7M$, and $L2_1 \rightarrow 5M \rightarrow 7M$ $\rightarrow L1_0$ with decreasing temperature. It is obvious that the degrees of chemical and structural disorder caused by quenching may change sufficiently the properties like total energies, magnetic moments and elastic moduli in both austenite and martensite phases.

In the present work, we focus on the theoretical investigation of the degrees of structural and chemical disorder in Mn-rich Ni-Mn-Ga alloys on their structural and magnetic properties.

Computational details

The crystal structure optimization of the systems studied were performed using the Vienna ab-initio Simulation Package (VASP) within the density functional theory [7]. To describe the exchange-correlation functional, the spin polarized generalized gradient approximation in the parameterization of Perdew-Burke-Ernzerhof was used [8]. A plane-wave cut-off of 750 eV and k-point grid of 8×8×8 were used. The *ab initio* calculations were performed for a 16-atom supercell, $Ni_8Mn_4Ga_4$. In order to model the off-stoichiometric chemical disorder in

Published under license by Materials Research Forum LLC.

Shape Memory Alloys – SMA 2018   Materials Research Forum LLC
Materials Research Proceedings **9** (2018) 118-121   doi: http://dx.doi.org/10.21741/9781644900017-22

$Ni_8Mn_{4+x}Ga_{1-x}$, the Ga atoms were replaced sequentially by Mn atoms. By this means the four compositions (x = 0, 1, 2 and 3) were taken into account. On the other hand, to model the structural disorder, the layered structure consisting of parallel rows of Mn and Ga atoms was considered. The ordered and disordered layered structures of $Ni_2Mn_{1+x}Ga_{1-x}$ are shown in Fig. 1.

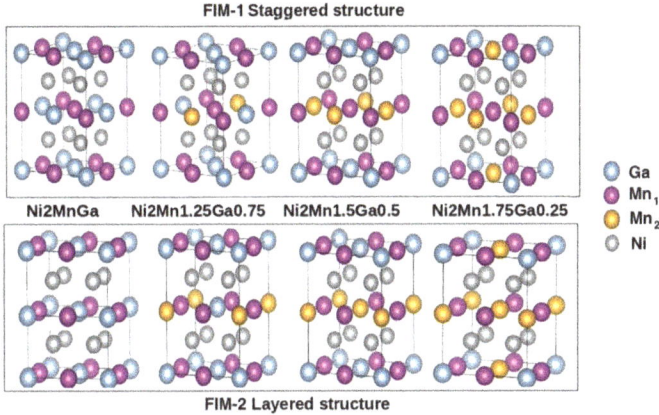

Figure 1. Tetragonal crystal structure of $Ni_2Mn_{1+x}Ga_{1-x}$ with staggered and layered disorder.

In the optimizations of the ordered and various disordered structures, we assumed the ferromagnetic (FM) order and two ferrimagnetic (FIM) orders labeled as FIM-1 ("staggered") and FIM-2 ("layered"). For the FIM states, Mn atoms which occupy Ga sublattice have reversed magnetic moment in contrast to the magnetic moments of Ni and Mn atoms located at their regular sublattices.

Results and discussion

The structural and magnetic properties of $Ni_2Mn_{1+x}Ga_{1-x}$ with staggered and layered disorder were calculated by the energy minimization procedure. The computed lattice parameters of cubic austenitic phase for studied compositions are given in Table 1 in their stable magnetic reference state along with available experimental lattice constants [6]. We would like to note that the experimental values were estimated assuming that the volume of crystal lattice change no practically due to austenite-martensite transformation ($a_0^3 \approx a_t^2 c$, where a_0 is the lattice constant of $L2_1$-cubic austenite and a_t and c are the lattice constants of $L1_0$-tetragonal martensite). It can be clearly seen that the optimized lattice parameter decreases slightly with increasing Mn content. Moreover, the calculated lattice constants are found to be in a good agreement with the experimental data.

We discuss next the possible stability of martensitic phase of Ni-Mn-Ga alloys with substitution of Mn atoms at Ga sites. To accomplish this, we performed the structural optimization calculations of tetragonally distorted cubic structure and compared the ground state energies of cubic austenite and tetragonal martensite phases in their corresponding magnetic reference states. In Fig. 2 we plot the total energy versus tetragonality for Mn-rich Ni-Mn-Ga alloys. In our calculations, we considered that the volume of unit cell is assumed to be constant during the tetragonal distortion.

Shape Memory Alloys – SMA 2018   Materials Research Forum LLC
Materials Research Proceedings **9** (2018) 118-121   doi: http://dx.doi.org/10.21741/9781644900017-22

Table 1. The computed values of the lattice parameter (in Å), magnetic reference state,
and total magnetic moment (in $\mu_B/f.u.$) of $Ni_2Mn_{1+x}Ga_{1-x}$ in the austenitic phase.
The experimental values of lattice parameters [6] are also quoted for comparison.

Composition	a_0	a_0 (exp)	Magnetic order	μ_{tot}
$x = 0$	5.811	5.822	FM	4.09
$x = 0.25$	5.809	5.814	FIM-1	3.08
$x = 0.5$	5.806	5.81	FIM-1	2.06
$x = 0.75$	5.804		FIM-1	0.99

We can find that in a case of the stoichiometric Ni_2MnGa, the global energy minimum for the energy curve with FM solution at $c/a \approx 1.25$ is clearly visible, suggesting that the martensitic transformation between $L2_1$ cubic and $L1_0$ tetragonal phase can occur in FM state only. Whereas the layered structure with FM and FIM-2 orders is not favorable. On the other side, an increase in the Mn content leads to the appearance of ferrimagnetic state in both austenite and martensite phases. We found that for composition with $x = 0.25$, the austenite-martensite transformation realizes in FIM-1 "staggered" state while with further increasing Mn content it occurs between austenite with FIM-1 "staggered" state and martensite with FIM-2 "layered" state.

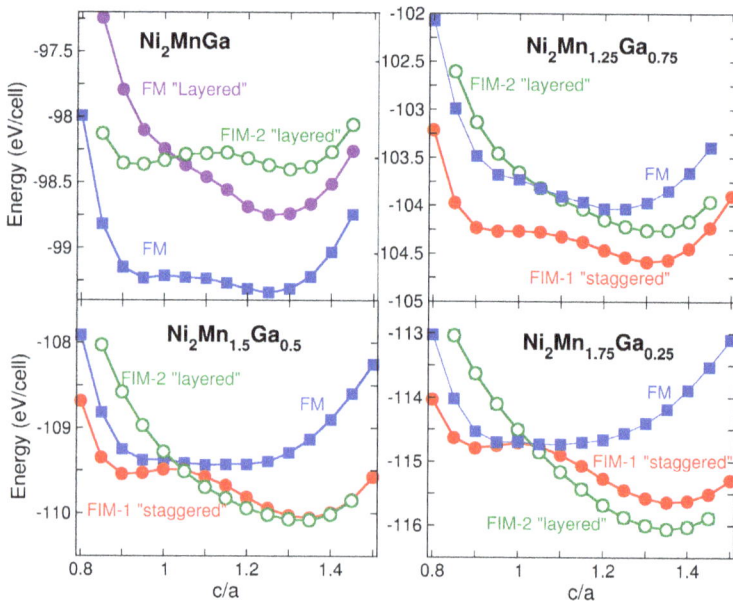

Figure 2. Calculated energy curves as a function of c/a ratio for $Ni_2Mn_{1+x}Ga_{1-x}$ in FM and FIM states. Here FM order in $L1_2$ structure and layered structure as well as FIM-1 ("staggered") and FIM-2 ("layered") spin configurations are considered.

Moreover, as can be seen that the total energy difference between the cubic austenite and the tetragonal martensite rises with substitution of Mn, implying the increase of martensitic transformation temperature (T_m) according to $E \approx k_B T_m$.

Summary

In summary, the magnetic and structural properties and phase stability of Heusler Mn-rich $Ni_2Mn_{1+x}Ga_{1-x}$ (x = 0, 0.25, 0.5, and 0.75) alloys are investigated by the first-principles calculations within the supercell approach. The geometric structure optimization has shown that the martensitic transformation appears between austenite and martensite phases with different magnetic ordering for all compositions considered. For compositions with x = 0 and 0.25, it occurs in FM and FIM-1 state, respectively. Whereas in a case of x = 0.5 and 0.75, it appears with a change in the magnetic order between the $L2_1$-ordered FIM-1 austenite and disordered layered FIM-2 martensite. Besides, it is shown that an increase in Mn content leads to an increase in the energy difference between austenite and martensite structure. This indicates on an increase in a martensitic transformation temperature in $Ni_2Mn_{1+x}Ga_{1-x}$ with increasing Mn concentration.

Acknowledgments

This work was supported by the Russian Foundation for Basic Research Grant 18-32-00507.

References

[1] A. A. Likhachev, A. Sozinov, K. Ullakko, Different modeling concepts of magnetic shape memory and their comparison with some experimental results obtained in Ni–Mn–Ga, Mater. Sci. Eng. A 378 (2004) 513-518. https://doi.org/10.1016/j.msea.2003.10.353

[2] A. Sozinov, A.A. Likhachev, K. Ullakko, Crystal structures and magnetic anisotropy properties of Ni-Mn-Ga martensitic phases with giant magnetic-field-induced strain, IEEE Trans. Magn. 38 (2002) 2814-2816. https://doi.org/10.1109/TMAG.2002.803567

[3] K. Ullakko, J.K. Huang, C. Kanter, V.V. Kokorin, R.C. O'Handley, Magnetic-field-induced strain in Ni₂MnGa shape-memory alloy, J. Appl. Phys. 81 (1998) 5416-1. https://doi.org/10.1063/1.364556

[4] V.V. Khovailo, K. Oikawa, T. Takagi, Entropy change at the martensitic transformation in ferromagnetic shapememory alloys $Ni_{2+x}Mn_{1-x}Ga$, J. Appl. Phys. 93 (2003) 8483-8485. https://doi.org/10.1063/1.1556218

[5] P.J. Webster, K.R.A. Ziebeck, S.L. Town, M.S. Peak, Magnetic order and phase transformation in Ni₂MnGa, Philos. Mag. B 49 (1984) 295-310. https://doi.org/10.1080/13642817408246515

[6] A. Çakir, L. Righi, F. Albertini, M. Acet, M. Farle, S. Akturk, Extended investigation of intermartensitic transitions in Ni–Mn–Ga magnetic shape memory alloys: a detailed phase diagram determination, J. Appl. Phys. 114 (2013) 183912-9. https://doi.org/10.1063/1.4831667

[7] G. Kresse, J. Furthmuller, Efficiency of ab initio total energy calculations for metals and semiconductors using a plane-wave basis set, Comput. Mater. Sci. 6 (1996) 15–50. https://doi.org/10.1016/0927-0256(96)00008-0

[8] J.P. Perdew, K. Burke, M. Ernzerhof, Generalized gradient approximation made simple, Phys. Rev. Lett. 77 (1996) 3865–3868. https://doi.org/10.1103/PhysRevLett.77.3865

Shape Memory Alloys – SMA 2018   Materials Research Forum LLC
Materials Research Proceedings 9 (2018) 122-127   doi: http://dx.doi.org/10.21741/9781644900017-23

The Effect of Pt-doping on Properties of Ni-Mn-(Ge, In) Heusler Alloys

Mikhail A. Zagrebin[1,2,a*], Vladimir V. Sokolovskiy[1,3,b], Vasiliy D. Buchelnikov[1,c],
Olga N. Miroshkina[1,d], Mariya V. Matyunina[1,e]

[1]Chelyabinsk State University, 129 Brat'ev Kashirinykh Str., Chelyabinsk 454001, Russia

[2]National Research South Ural State University, 76 Lenin Prospect, Chelyabinsk 454080, Russia

[3]National University of Science and Technology "MISiS", 4 Leninskiy Prospect, Moscow 119049, Russia

[a]miczag@mail.ru, [b]vsokolovsky84@mail.ru, [c]buche@csu.ru, [d]miroshkina.on@yandex.ru, [e]matunins.fam@mail.ru

*corresponding author

Keywords: Density Functional Theory, Exchange Interaction, Curie Temperature, Heusler Alloys

Abstract. In this paper, structural (lattice parameter, tetragonal distortions) and magnetic (total magnetic moments, magnetic exchange parameters) properties for $Ni_{2-x}Pt_xMn(Ge,In)$ Heusler alloys were studied using density functional theory. It was shown that the equilibrium crystal lattice parameter of the austenite increases with increasing Pt content. The investigation of the magnetic exchange parameters shows that the largest contribution to the total exchange energy is associated with the interaction between the nearest neighboring Ni-Mn atoms. The concentration dependences of the Curie temperature were calculated by means Monte Carlo method. The temperatures of the martensitic phase transitions were estimated. It was shown that the temperature of martensitic phase transition increases with increasing Pt content. Based on these temperatures, a phase diagram for $Ni_{2-x}Pt_xMn(Ge, In)$ Heusler alloys for the whole range of Pt concentration ($0 \leq x \leq 2$) is plotted.

Introduction

Ni-Mn-Z (Z = Ga, In, Sn) Heusler alloys are of interesting for practical applications due to the numerous unique effects such as shape memory effect in the ferromagnetic state, magnetic shape memory effect, giant magnetocaloric effect, magnetostriction, etc. The strong magnetoelastic interaction between the magnetic and structural subsystem is one of the sources of these effects. The magnetic-field-induced strain in Ni-Mn-Ga alloys reaches about 10 % [1-3]. However, recent *ab initio* works have shown that the value of the magnetic-field-induced strain can achieve 14 % by Pt-doping in this alloy [4,5]. The results of structural and magnetic properties of Ni-Pt-Mn-Z (Z = Ge, In) Heusler alloys by means of *ab initio* methods are presented.

Calculation details

To study the ground states and magnetic properties of Heusler $Ni_{2-x}Pt_xMn(Ge, In)$ ($0 \leq x \leq 2$) alloys, the projector augmented wave (PAW) and Korringa-Kohn-Rostoker methods were used. These methods are realized in VASP (Vienna *ab initio* simulation program) [6,7] and SPRKKR (spin-polarized-relativistic Korringa-Kohn-Rostoker code) [8] computational packages, respectively. To take into account the exchange-correlation interaction, the General gradient approximation (GGA) in the Perdew-Burke-Ernzerhof (PBE) parameterization was used [9]. The geometrical structure optimization of the austenitic phase was carried out using VASP package in combination with a supercell approach. 16-atom supercell with $L2_1$ structure (space

Published under license by Materials Research Forum LLC.

Shape Memory Alloys – SMA 2018   Materials Research Forum LLC
Materials Research Proceedings **9** (2018) 122-127   doi: http://dx.doi.org/10.21741/9781644900017-23

symmetry group $Fm\bar{3}m$, No. 225) was used for geometrical optimization. This structure follows from experimental studies of Ni-Mn-Z Heusler alloys. It consists of four interpenetrating fcc sublattices: two sublattices of X atoms that are located at the $8c$ (positions (1/4, 1/4, 1/4) and (3/4, 3/4, 3/4)) Wyckoff position and the sublattices of the Z and Y atoms, which occupy $4a$ and $4b$ ((0, 0, 0) and (1/2, 1/2, 1/2)) Wyckoff position, respectively. Additional Pt atoms take positions of Ni. To estimate the optimized lattice constants of all alloys under study, the Birch-Murnaghan equation of state was used. The optimized lattice parameters were used to calculate the magnetic moments and exchange coupling constants (J_{ij}) with the help of the SPR-KKR package. The magnetic moments and magnetic exchange constants were calculated using the GGA-PBE parameterization. All calculations were carried out for the ferromagnetic (FM) configuration in which all atoms have positive magnetic moments. The obtained long-range exchange coupling constants allow us to simulate the temperature dependences of magnetization in the framework of the classical three-dimensional Heisenberg model ($H = -\sum_{ij} J_{ij}\mathbf{S}_i\mathbf{S}_j$) and

Monte Carlo (MC) routine [10].

Results and discussion
At the first stage, to determine the equilibrium lattice parameters of $Ni_{2-x}Pt_xMn(Ge, In)$ alloys, the dependencies of the total energy on the lattice parameter a were calculated. For the stoichiometric Ni_2MnGe and Ni_2MnIn alloys $a_0 = 5.81$ Å and 6.07 Å, respectively, for the Pt_2MnGe (Pt_2MnIn) alloys $a_0 = 6.28$ (6.44) Å. It was shown that the equilibrium crystal lattice parameters a_0 of the austenite increase with increasing Pt concentration for both alloys. The cause of this behavior is the greater atomic radius of Pt atoms than the radius of Ni. This behavior is in good agreement with published experimental and theoretical data obtained for $Ni_{2-x}Pt_xMnGa$ Heusler alloys [11].

In order to investigate the possibility of martensitic transformation in these alloys, the total energies as a function of the tetragonal distortions of the cubic structure along the z axis were calculated. It is well known that the austenite-martensite transition requires the martensitic phase has a lower energy than austenitic one. The total energy differences between the tetragonal distorted and cubic phases $\Delta E = E - E_0$ for $Ni_{2-x}Pt_xMnGe$ and $Ni_{2-x}Pt_xMnIn$ alloys as functions of c/a ratio are depicted in Fig. 1(a) and (b). In this case, the zero value of ΔE corresponds to the austenitic phase.

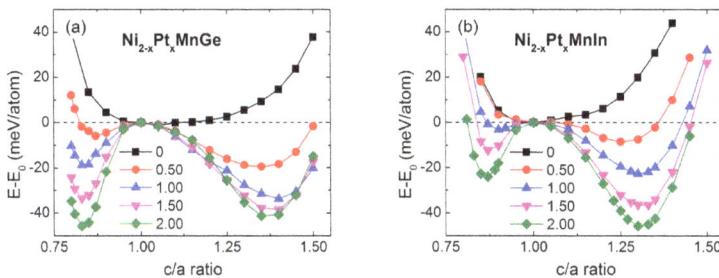

Figure 1. The total energy differences for (a) $Ni_{2-x}Pt_xMnGe$ and (b) $Ni_{2-x}Pt_xMnIn$ as functions of a tetragonal distortion (c/a).

Shape Memory Alloys – SMA 2018   Materials Research Forum LLC
Materials Research Proceedings 9 (2018) 122-127   doi: http://dx.doi.org/10.21741/9781644900017-23

The calculations show that in the case of Ni_2MnGe and Ni_2MnIn alloys austenite-martensite transformation is not realized due to the absence of energy minimum in the $\Delta E(c/a)$ curve in the martensitic phase. The addition of Pt atoms into Ni-Mn(Ge, In) systems and an increase of their content result in the appearance of stable martensitic phase at $c/a > 1.20$. Besides, another minimum in the energy curves at $c/a < 1$ which indicate the metastable martensitic phase can be found for compositions with Pt concentration $x \geq 0.25$ for Ni-Pt-Mn-Ge and $x \geq 0.75$, respectively. Moreover, the energy minimum at $c/a = 0.84$ has the lowest energy for Pt_2MnGe. The energy difference ΔE and magnitude of tetragonality c/a increase (excepting Pt_2MnGe) with increasing Pt concentration (x). This is important since this difference is a quantitative indicator of the martensitic transformation temperature (T_m) in accordance with the ratio $\Delta E \approx k_B T_m$, where k_B is the Boltzmann constant [5]. The calculated lattice parameters, tetragonal distortion c/a, total energy difference ΔE of $Ni_{2-x}Pt_xMn(Ge, In)$ Heusler alloys are summarized in Table 1. We note that calculated temperatures T_m will be used for the construction of the phase diagram.

Table 1. Calculated equilibrium lattice parameters a_0 (in Å), tetragonal distortion c/a and the total energy difference ΔE (in meV/atom) for $Ni_{2-x}Pt_xMn(Ge, In)$ Heusler alloys.

x	0.00	0.25	0.50	0.75	1.00	1.25	1.50	1.75	2.00
				$Ni_{2-x}Pt_xMnGe$					
a_0	5.81	5.89	5.96	6.02	6.07	6.13	6.19	6.24	6.28
c/a	–	1.29	1.35	1.40	1.40	1.40	1.39	1.38	0.84
ΔE	–	–0.14	–19.49	–26.20	–33.88	–38.26	–38,41	–41.33	–45.90
				$Ni_{2-x}Pt_xMnIn$					
a_0	6.07	6.14	6.19	6.24	6.28	6.33	6.37	6.40	6.44
c/a	–	1.21	1.25	1.29	1.305	1.31	1.31	1.31	1.31
ΔE	–	–1.30	–8.47	–16.15	–22.85	–30.77	–37.44	–43.09	–48.58

At the second stage, *ab initio* calculations of the magnetic properties in the austenite state were performed for $Ni_{2-x}Pt_xMn(Ge, In)$ $(0 \leq x \leq 2)$ Heusler alloys. The magnetic properties allow to predict compositional trends of the Curie temperature; thus, they are of significant interest. Calculations of magnetic properties were performed for the cell consisting of four atoms by using the SPR-KKR package and coherent approximation approach. The calculated total and partial spin magnetic moments are summarized in Table 2.

Table 2. Calculated total and partial spin magnetic moments (in μ_B) of $Ni_{2-x}Pt_xMn(Ge, In)$ Heusler alloys.

x	0.00	0.25	0.50	0.75	1.00	1.25	1.50	1.75	2.00
				$Ni_{2-x}Pt_xMnGe$					
total	3.89	3.98	4.04	4.07	4.10	4.17	4.21	4.21	4.19
μ_{Ni}	0.20	0.22	0.23	0.24	0.25	0.27	0.29	0.30	–
μ_{Pt}	–	0.09	0.09	0.09	0.10	0.10	0.10	0.11	0.10
μ_{Mn}	3.55	3.63	3.69	3.75	3.79	3.85	3.91	3.95	3.99
μ_{Ge}	–0.06	–0.05	–0.05	–0.04	–0.03	–0.01	0.00	0.00	0.00
				$Ni_{2-x}Pt_xMnIn$					
total	4.02	4.08	4.11	4.14	4.15	4.15	4.14	4.12	4.10
μ_{Ni}	0.29	0.31	0.32	0.34	0.36	0.37	0.39	0.40	–
μ_{Pt}	–	0.10	0.10	0.10	0.11	0.11	0.11	0.11	0.11
μ_{Mn}	3.55	3.61	3.66	3.71	3.76	3.80	3.83	3.87	3.90
μ_{In}	–0.10	–0.09	–0.09	–0.08	–0.07	–0.06	–0.05	–0.04	–0.03

Shape Memory Alloys – SMA 2018   Materials Research Forum LLC
Materials Research Proceedings **9** (2018) 122-127   doi: http://dx.doi.org/10.21741/9781644900017-23

Conducted first-principles calculations have shown that the magnetic moment of Ni_2MnGe is smaller in comparison with the magnetic moment of Ni_2MnIn. Total magnetic moments increase with increasing of Pt concentration while $x < 1.25$ (1.50 in case of $Ni_{2-x}Pt_xMnGe$) and decrease after that. Partial magnetic moments increase with increasing of Pt concentration (x). Partial moments of Ge and In are negative and slightly increase with increasing of Pt excess (x). Partial magnetic moments of Pt are also slightly increasing.

The magnetic exchange interaction of the $Ni_{2-x}Pt_xMn(Ge, In)$ alloys as a function of the distance between the interacting pairs of atoms for the austenite is shown in Fig. 2. Positive values of the exchange constants ($J_{ij} > 0$) indicate the FM interaction, whereas the negative values ($J_{ij} < 0$) indicate an antiferromagnetic (AFM) interaction. We note that exchange interactions involving Ga and Pt atoms (except Mn-Pt coupling) are very small and are not presented in the figures.

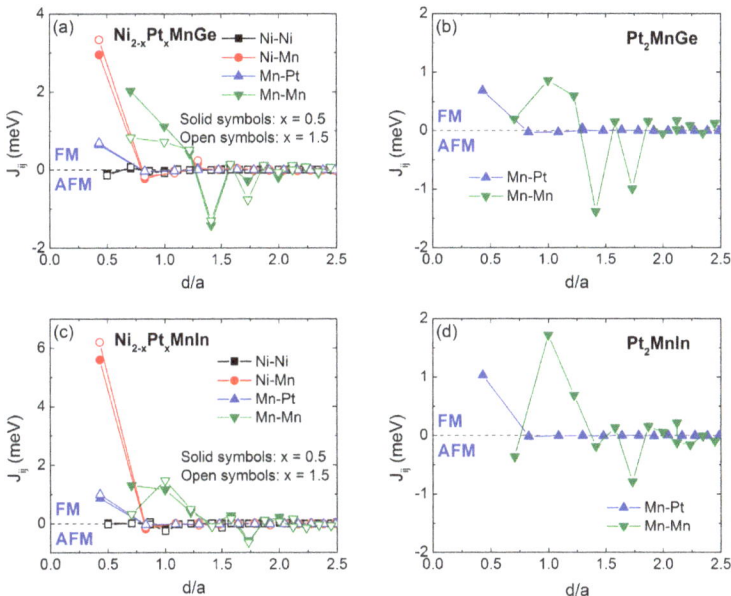

Figure 2. Exchange coupling constants as a function of distance (d/a) between i and j atoms for (a) $Ni_{2-x}Pt_xMnGe$ (x = 0.5, 1.5), (b) Pt_2MnGe, (c) $Ni_{2-x}Pt_xMnIn$ (x = 0.5, 1.5), and (d) Pt_2MnIn alloys in the austenitic states.

It can be seen from Fig. 2 that the magnetic exchange interactions constants exhibit a damping oscillatory behavior for all compositions. The magnetic interactions between Mn-Ni, Mn-Pt pairs are FM and slightly decrease with increasing distance between atoms. Mn-Mn interactions are also FM predominantly, but they can have an AFM contribution (beginning from fourth coordination shell) depending on the distance between Mn atoms. Interactions between the nearest pairs of Ni(Pt)-Mn atoms provide the largest contribution to exchange in comparison with intra-sublattice interactions (Ni-Ni and Mn-Mn). Note, that in case of Ni-Pt-Mn-In alloys Ni-Mn interaction in the first coordinational shell is larger in two times in comparison with Ni-Pt-Mn-Ge alloys. Mn-Pt interaction approximately equal for both series of alloys. The

Shape Memory Alloys – SMA 2018   Materials Research Forum LLC
Materials Research Proceedings **9** (2018) 122-127   doi: http://dx.doi.org/10.21741/9781644900017-23

magnitude of the interaction between nearest neighbors Ni-Mn increases with increasing Pt concentration. At the same time, the interaction between Mn-Mn in the first coordination sphere is FM and it decreases with increasing of Pt concentration. In the case of Pt_2MnIn this interaction in the first coordination sphere becomes a weak AFM. The interactions of Ni-Ni and Mn-Pt have a weak dependence on the Pt concentration. Calculated constants allow us to estimate the Curie temperature for cubic $Ni_{2-x}Pt_xMn(Ge, In)$ alloys in the framework of MC simulations. The ferro-paramagnetic phase transition in the austenitic phase occurs during heating for all studied alloys. Obtained values of Curie temperatures (together with estimated martensitic transition temperatures) has been plotted on the T–x phase diagrams for $Ni_{2-x}Pt_xMnGe$ and $Ni_{2-x}Pt_xMnIn$ $(0 \leq x \leq 2)$ alloys, as shown in Fig. 3.

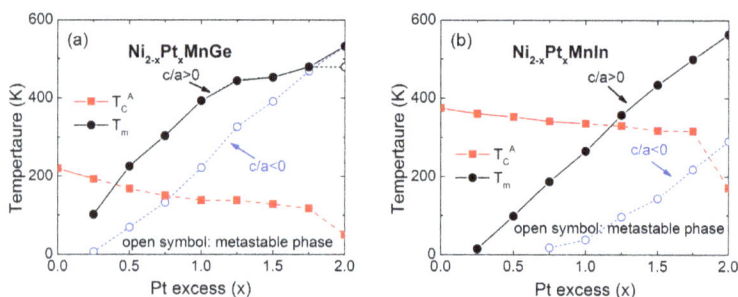

Figure 3. Calculated Curie temperatures (T_C) and estimated temperatures of the martensitic phase transition (T_m) for (a) $Ni_{2-x}Pt_xMnGe$ and (b) $Ni_{2-x}Pt_xMnIn$ ($0 \leq x \leq 2$) alloys as functions of Pt concentration. Open symbols denote temperature of transition to the metastable martensitic phase.

Here, solid lines depict real Curie temperatures, while dashed lines with symbols represent "virtual" temperatures, which is responsible for a hypothetical magnetic transition at significantly higher temperatures than the temperature of the martensitic transformation [9]. It is seen from the phase diagram that the Curie temperatures of austenite (T_C^A) decrease, while the temperatures of martensitic phase transition (T_m) increase with increasing Pt content.

Summary
Structural and magnetic properties of $Ni_{2-x}Pt_xMn(Ge, In)$ alloys have been investigated with the help of the combination of *ab initio* and MC calculations. It is shown that the equilibrium crystal lattice parameter of the austenite increases with increasing Pt content. An investigation of the magnetic exchange parameters shows that the largest contribution to the total exchange energy is associated with the interaction between nearest neighboring Ni-Mn atoms. The concentration dependences of the Curie temperature were calculated by means Monte Carlo method. Based on estimated temperatures of the martensitic phase transitions, a phase diagram for $Ni_{2-x}Pt_xMn(Ge, In)$ Heusler alloys for the whole range of Pt concentration ($0 \leq x \leq 2$) is constructed. It can be seen from this phase diagrams that the Curie temperatures of austenite (T_C^A) decrease, while the temperatures of the martensitic phase transition (T_m) increase with increasing Pt content.

Acknowledgments
This work was supported by Russian Science Foundation (17-72-20022).

Shape Memory Alloys – SMA 2018   Materials Research Forum LLC
Materials Research Proceedings **9** (2018) 122-127   doi: http://dx.doi.org/10.21741/9781644900017-23

References

[1] K. Ullakko, J.K. Huang, C. Kantner, R.C. O'Handley, V. V. Kokorin, Large magnetic - field - induced strains in Ni₂MnGa single crystals, Appl. Phys. Lett. 69 (1996) 1966-1968. https://doi.org/10.1063/1.117637

[2] S.J. Murray, M. Marioni, S.M. Allen, R.C. O'Handley, T.A. Lograsso, 6% magnetic-field-induced strain by twin-boundary motion in ferromagnetic Ni-Mn-Ga, Appl. Phys. Lett.77 (2000) 886-888. https://doi.org/10.1063/1.1306635

[3] A. Sozinov, A.A. Likhachev, N. Lanska, K. Ullakko, Giant magnetic-field-induced strain in NiMnGa seven-layered martensitic phase, Appl. Phys. Lett. 80 (2002) 1746-1748. https://doi.org/10.1063/1.1458075

[4] M. Siewert, M.E. Gruner, A. Hucht, H.C. Herper, A. Dannenberg, A. Chakrabarti, N. Singh, R. Arróyave, P. Entel, A First-Principles Investigation of the Compositional Dependent Properties of Magnetic Shape Memory Heusler Alloys, Adv. Eng. Mat. 14 (2012) 530-546. https://doi.org/10.1002/adem.201200063

[5] P. Entel, M. Siewert, M.E. Gruner, A. Chakrabarti, S.R. Barman, V.V. Sokolovskiy, V.D. Buchelnikov, Optimization of smart Heusler alloys from first principles, J. Alloy. Compd. 577 (2013) S107-S112. https://doi.org/10.1016/j.jallcom.2012.03.005

[6] G. Kresse, J. Furthmuller, Efficient iterative schemes for ab initio total-energy calculations using a plane-wave basis set, Phys. Rev. B. 54 (1996) 11169-11186. https://doi.org/10.1103/PhysRevB.54.11169

[7] G. Kresse, D. Joubert, From ultrasoft pseudopotentials to the projector augmented-wave method, Phys. Rev. B. 59 (1999) 1758-1775. https://doi.org/10.1103/PhysRevB.59.1758

[8] H. Ebert, D. Ködderitzsch, J. Minár, Calculating condensed matter properties using the KKR-Green's function method – Recent developments and applications, Rep. Prog. Phys. 74 (2011) 096501. https://doi.org/10.1088/0034-4885/74/9/096501

[9] J.P. Perdew, K. Burke, M. Ernzerhof, Generalized Gradient Approximation Made Simple, Phys. Rev. Lett. 77 (1996) 3865-3868. https://doi.org/10.1103/PhysRevLett.77.3865

[10] D.P. Landau, K. Binder, A Guide to Monte Carlo Simulations in Statistical Physics, Cambridge University Press, Cambridge, 2005. https://doi.org/10.1017/CBO9780511614460

[11] M.A. Zagrebin, S.A. Derevyanko, V.V. Sokolovskiy, V.D. Buchelnikov, Complex investigations of phase diagram of Ni-Pt-Mn-Ga Heusler alloys, Letters on Materials 8 (2018) 21-26. https://doi.org/10.22226/2410-3535-2018-1-21-26

Materials Research Forum LLC

doi: http://dx.doi.org/10.21741/9781644900017

Novel Materials: Design, Synthesis, Functional Properties

Shape Memory Alloys – SMA 2018
Materials Research Proceedings 9 (2018) 131-135

Materials Research Forum LLC
doi: http://dx.doi.org/10.21741/9781644900017-24

Multicaloric Effect and Magnetoelectric Coupling in Fe$_{48}$Rh$_{52}$/PZT Composite

Ivan A. Starkov[1,2,a], Abdulkarim A. Amirov[3,4,b*], Alexander S. Starkov[2,c]

[1]Nanotechnology Center, St. Petersburg Academic University, 8/3 Khlopin Str., St. Petersburg 194021, Russia

[2]National Research University of Information Technologies, Mechanics and Optics (ITMO University), 49 Kronverksky Prospect, St. Petersburg 197101, Russia.

[3]Laboratory of Novel Magnetic Materials & Institute of Physics Mathematics and Informational Technologies Immanuel Kant Baltic Federal University, 14 A. Nevskogo Str., Kaliningrad 236041, Russia

[4]Amirkhanov Institute of Physics, Daghestan Scientific Center, Russian Academy of Sciences, 94 Yagarskogo Str., Makhachkala 367003, Russia

[a]starkov@spbau.ru, [b]amiroff_a@mail.ru, [c]ferroelectrics@ya.ru

*corresponding author

Keywords: Multiferroics, Composites, Magnetoelectric Effect, Magnetocaloric Effect, Multicaloric Effect

Abstract. The multicaloric effect on the example of Fe$_{48}$Rh$_{52}$/PZT multiferroic composite was theoretically studied based on experimental results. The interrelation between multicaloric and magnetoelectric effect around metamagnetic phase transition in 315 K was observed. The calculations were carried out using generalized matrix averaging method.

Introduction

Recently the studies of materials with giant caloric effects (CE) of various nature has significantly increased due to their practical applicability in alternative traditional, energy-efficient and environmentally safe cooling systems using solid-state compounds as an element base. The originally known magnetocaloric (MCE), electrocaloric (ECE), barocaloric and elastocaloric effects [1-3], which concludes in temperature and entropy changes under external magnetic, electric or elastic fields changes. The idea of practical usage of coexisting CEs is the basis of one of the modern and rapidly direction in condensed matter physics. Approach of simultaneous observation of at least two of the well-known CEs for research of thermodynamic material properties is called "multicaloric effect" and the materials are imprecisely called "multicalorics" [4].

As known, multiferroics are materials that demonstrate the simultaneous coexistence of at least two known ferro ordering (magnetic, electric, mechanical). It is a reason for considering multiferroics as potential candidates for observing multicaloric effects in them. The multicaloric effect theoretically and experimentally was studied in [5-9] and obtained results observe small magnitudes of multicaloric effect in natural multiferroics. Additionally it was demonstrated that CEs in multiferroics closely depend from magnetoelectric (ME) interrelation [8,9]. From this point of view, the magnetoelectric composites were proposed for multicaloric studies, due to their strong ME coupling [10-11]. In present paper on the base of experimental results reported in [12], the multicaloric effect in Fe$_{48}$Rh$_{52}$/PZT magnetoelectric layered composite was studied. The bilayered ME composite was consist of magnetocaloric Fe$_{48}$Rh$_{52}$ layer and piezoelectric layer of PbZr$_{0.53}$Ti$_{0.47}$O$_3$ (PZT). The ME coupling is induced as results of mechanical interaction between layers due to the magnetostriction and piezoelectric effect. The generalized matrix

Published under license by Materials Research Forum LLC.

Shape Memory Alloys – SMA 2018   Materials Research Forum LLC
Materials Research Proceedings **9** (2018) 131-135   doi: http://dx.doi.org/10.21741/9781644900017-24

averaging (GMA) method for calculating the parameters of an effective medium with physical properties equivalent to those of a set of thin multiferroic layers was used in the framework theoretical consideration of multicaloric effect [10,11].

Multicaloric effect in Fe$_{48}$Rh$_{52}$/PZT multiferroics

Since the temperature change in the composite depends on two fields, magnetic H and electric E, we must understand that we are dealing with the multicaloric effect (μCE). For the theoretical description of μCE, we can introduce the free energy F in the simplest form [11]

$$F = a_m \frac{M^2}{2} + a_e \frac{P^2}{2} + \alpha PM - EP - HM. \tag{1}$$

Here, M is the magnetization and P is the polarization. The coefficient a_m (a_e) has the meaning of the inverse magnetic (dielectric) permittivity for the composite and α is the theoretical magnetoelectric coefficient. According to our estimates, the characteristic relaxation time of thermal processes for the Fe$_{48}$Rh$_{52}$ composite is no more than 3×10^{-3} s. Therefore, we can use equilibrium thermodynamics to describe ongoing thermal processes. The conditions of minimality of F lead to the relations $H = a_m M + \alpha P$, $E = \alpha M + a_m P$, or

$$M = \frac{a_e H - \alpha E}{a_e a_m - \alpha^2}, \quad P = \frac{a_m E - \alpha H}{a_e a_m - \alpha^2}. \tag{2}$$

The factor in front of H in Eq. 2 is the magnetoelectric susceptibility coefficient, which is defined as $\alpha_{ME} = dP/dH$. The relationship between the coefficients α_{ME} and α_E is given by $\alpha_{ME} = \varepsilon_0 \varepsilon \alpha_E$, with ε and ε_0 as the permittivity of free space and relative permittivity of the material, respectively.

This relationship is required for the comparison of experimental and theoretical results. On the basis of the formulas obtained, it is possible to find the caloric quantities. The first is the entropy

$$S = \frac{\partial a_m}{\partial T} \frac{M^2}{2} + \frac{\partial a_e}{\partial T} \frac{P^2}{2} + \frac{\partial \alpha}{\partial T} PM. \tag{3}$$

The three terms in the entropy equation are necessary for the existence of three caloric effects: magnetocaloric, electrocaloric (ECE), and magnetoelectrocaloric (MECE). In accordance with this, the adiabatic change in temperature is represented in the form

$$dT_{\mu CE} = dT_{MCE} + dT_{ECE} + dT_{MECE}. \tag{4}$$

For the reason that the phase transition in PZT occurs far from the considered temperature range, the ECE can be neglected. In addition, as the MCE has been studied in detail [1], we write out only expressions for MECE

$$T_{MECE} = -\frac{T}{C} \frac{\partial \alpha}{\partial T} \left(\frac{\partial PM}{\partial E} dE + \frac{\partial PM}{\partial H} dH \right). \tag{5}$$

Here, C is the heat capacity of the composite. The derivatives on the right-hand side of Eq. 5 are easily found using Eq. 2. For small values of the magnetoelectric coefficient α, the derived relations can be simplified. By neglecting α^2 in the denominator of Eq. 2, we conclude that all three introduced magnetoelectric coefficients α_{ME}, α_e and α are proportional to each other. The

Shape Memory Alloys – SMA 2018   Materials Research Forum LLC
Materials Research Proceedings 9 (2018) 131-135   doi: http://dx.doi.org/10.21741/9781644900017-24

coefficients of proportionality can be considered independent of temperature due to the smallness of the total temperature change. As a result, we get that

$$dT_{MECE} \sim \frac{\partial \, \alpha_{ME}}{\partial T}. \tag{6}$$

On the basis of this formula, the temperature dependence of MECE is built in Fig. 1. The same figure shows the experimental MECE dependence obtained from the relation Eq. 4. The existing discrepancy in the shape of theoretical and experimental curves (temperature shift) is explained by the fact that in the theoretical graph the quantity α_{ME} is taken in the absence of the electric field, as it was necessary.

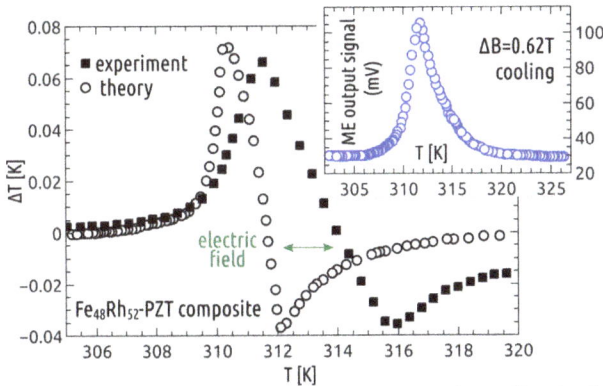

Figure 1. The temperature dependence of the magnetoelectrocaloric effect for the $Fe_{48}Rh_{52}$/PZT bilayer. Inset: the temperature dependence of the magnetoelectric coefficient. The amplitude of the AC magnetic field is $\Delta B = 0.62$ T and frequency $f = 3Hz$.

Magnetoelectric effect in $Fe_{48}Rh_{52}$/PZT multiferroics

The inset of Fig. 1 shows the ME coefficient as a function of temperature for the $Fe_{48}Rh_{52}$/PZT composite. The output voltage exhibits the maximum near AFM-FM transition temperature in an AC magnetic field of amplitude 0.62 T and frequency of 3Hz. The maximum of the magnetoelectric coefficient close to the AFM-FM transition temperature corresponds to the large magnetic field-induced mechanical strain (magnetostriction). These results are in a good agreement with the model proposed in [13], where the correlation between the ME effect and magnetic-field-induced strain of magnetic layer was theoretically and experimentally demonstrated. According to the model described in [13,14], the magnetoelectric response of bilayer composite is proportional to the ME coefficient α_{ME} and is given by

$$\alpha_{ME} = \frac{2t(1-t)d_{31}q_{31}\mu_0\overline{s}}{(2td_{31}^2 - \varepsilon^p\overline{s})\left[\overline{\mu s} - 2q_{31}^2(1-t)^2\right]}, \tag{7}$$

where $\overline{s} = t\left(s_{11}^p + s_{12}^p\right) + (1-t)\left(s_1^m + s_{12}^m\right)$ and $\overline{\mu} = t + (1-t)\mu^m$. The subscripts m and p refer to the magnetostrictive and piezoelectric phases, respectively; d and q are the piezoelectric and piezomagnetic cooling coefficients, s is the compliance coefficient; μ_0 and μ are, respectively, permeabilities of the free space and magnetic phase; t is the fractional thickness for the piezoelectric layer. From the above equation, it becomes clear that α_{ME} of the composite structure depends on the ratio of the thicknesses of the layers as well as on the magnetic/magnetostrictive/elastic coefficients of the $Fe_{48}Rh_{52}$ layer and electric/piezomagnetic/elastic properties of the PZT layer.

Conclusions

The results of the magnetic, magnetoelectric, and magnetocaloric investigations demonstrate the influence of the electric field on themagnetocaloric effect of the layered $Fe_{48}Rh_{52}$/PZT multiferroic composite and large magnetoelectric ordering around the temperature of the metamagnetic phase transition. The presented experimental data is well described by the existing theoretical models for magneto- and multicaloric effects. The demonstrated results can be used for modeling of ME composite systems for studies of caloric effects.

Acknowledgements

The contribution of A.A. Amirov was funded by RFBR, according to the research project No.18-32-01036 mol_a. A.S. Starkov was supported by RSCF, research project No.18-19-00512.

References

[1] A.M. Tishin, Y.I. Spichkin, The Magnetocaloric Effect and its Applications, Bristol and Philadelphia, Institute of Physics Publishing, 2003. https://doi.org/10.1887/0750309229

[2] G. Suchaneck, V. Pakhomov, G. Gerlach, Electrocaloric Cooling, Chapter 2 in book: Refrigeration: Orhan Ekren, IntechOpen, 2017.

[3] A. Chauhan, S. Patel, R. Vaish, C.R. Bowen, A review and analysis of the elasto-caloric effect for solid-state refrigeration devices: Challenges and opportunities, Mater. Res. Soc. 2 (2015) 1-18. https://doi.org/10.1557/mre.2015.17

[4] L. Manosa, D. Gonzalez-Alonso, A. Planes, E. Bonnot, M. Barrio, J.L. Tamarit, S. Aksoy, M. Acet, Giant solid-state barocaloric effect in the Ni–Mn–In magnetic shape-memory alloy, Nat. Mater. 9 (2010) 478-481. https://doi.org/10.1038/nmat2731

[5] A. Planes, T. Castan, A. Saxena, Thermodynamics of multicaloric effects in multiferroics, Philos. Mag. 94 (2014) 1893-1908. https://doi.org/10.1080/14786435.2014.899438

[6] S. Fahler, U.K. Robler, O. Kastner, J. Eckert, G. Eggeler, H. Emmerich, P. Entel, S. Muller, E. Quandt, K. Albe, Caloric Effects in Ferroic Materials: New Concepts for Cooling, Adv. Eng. Mater. 14 (2012) 10-19. https://doi.org/10.1002/adem.201100178

[7] I. Flerov, E. Mikhaleva, M. Gorev, A. Kartashev, Caloric and multicaloric effects in oxygen ferroics and multiferroics, Phys. Solid State. 57 (2015) 429–441. https://doi.org/10.1134/S1063783415030075

[8] A. Starkov, I. Starkov, Multicaloric effect in a solid: New aspects, J. Exp. Theor. Phys. 119 (2014) 258-263. https://doi.org/10.1134/S1063776114070097

[9] M.M. Vopson, The multicaloric effect in multiferroic materials, Solid State Commun. 152 (2012) 2067-2070. https://doi.org/10.1016/j.ssc.2012.08.016

[10] A.S. Starkov, I.A. Starkov, Application of a generalized matrix averaging method for the calculation of the effective properties of thin multiferroic layers theoretical model for thin ferroelectric films and the multilayer structures based on them, J. Exp. Theor. Phys. 119 (2014) 861-869. https://doi.org/10.1134/S1063776114110120

[11] I.A. Starkov, A.S. Starkov, Effective parameters of multilayered thermo-electro-magneto-elastic solids, Solid State Communications 226 (2016) 5-7. https://doi.org/10.1016/j.ssc.2015.11.002

[12] A.A. Amirov, V.V. Rodionov, I.A. Starkov, A.S. Starkov, A.M. Aliev, J. Magn. Magn. Mater., In Press, Corrected Proof (2018). DOI: 10.1016/j.jmmm.2018.02.064. https://doi.org/10.1016/j.jmmm.2018.02.064

[13] Min Zeng, Siu Wing Or, Helen Lai Wa Chan, Large magnetoelectric effect from mechanically mediated magnetic field-induced strain effect in Ni-Mn-Ga single crystal and piezoelectric effect in PVDF polymer, J. Alloys Comp. 490 (2010) 5-8. https://doi.org/10.1016/j.jallcom.2009.09.167

[14] Ce-Wen Nan, M.I. Bichurin, Shuxiang Dong, D. Viehland, G. Srinivasan, Multiferroic magnetoelectric composites: Historical perspective, status, and future directions, J. Appl. Phys. 103, (2008) 031101-1-031101-35. https://doi.org/10.1063/1.2836410

Shape Memory Alloys – SMA 2018   Materials Research Forum LLC
Materials Research Proceedings 9 (2018) 136-139   doi: http://dx.doi.org/10.21741/9781644900017-25

Calculation of Effective Permittivity and Permeability for Iron Ore – Biochar – Bentonite Binder Powders Mixture

Anton P. Anzulevich[1,a*], Igor V. Bychkov[1,2,b], Vasiliy D. Buchelnikov[1,c],
Svetlana N. Anzulevich[1,d], Dmitry A. Kalganov[1,e], Zhiwei Peng[3,f]

[1]Chelyabinsk State University, 129 Brat'ev Kashirinykh Str., Chelyabinsk 454001, Russia

[2]National Research South Ural State University, 76 Lenin Prospect, Chelyabinsk 454080, Russia

[3]School of Minerals Processing and Bioengineering, Central South University, Changsha, Hunan 410083, China

[a]anzul@list.ru, [b]bychkov@csu.ru, [c]buche@csu.ru, [d]anzulic@gmail.com, [e]kalganov@csu.ru, [f]zwpeng@csu.edu.cn

*corresponding author

Keywords: Core-Shell Particles, Effective Medium Approximation, Biochar, Iron Reduction, Effective Permittivity, Effective Permeability

Abstract. A model of biochar – iron ore – binder powders mixture was studied at microwaves. The effective medium Bruggeman expression was expanded for three types of particles and for two types of core-shell particles. Surface impedance dependency on the volume fraction of iron ore was calculated.

Introduction

The mixture of biochar, iron ore, and binder (SiO_2 and Al_2O_3) is interesting in terms of use as a source material in steel-making industry. By using biochar to replace coal and coke, faster reduction and lower reduction temperature can be achieved with much lower emission of hazardous gas (SO_2) and CO_2. Then, the metallized pellet can be used as a good raw material for steel-making in electric arc furnace. Many publications studying the possibilities of its application in various fields of science and technology can be observed [1-3]. To achieve better performance of iron reduction from the considered mixture, it is required to improve impedance matching depending on volume fractions of components. Thus, in this paper, we simulate effective electrodynamic parameters of source mixture using effective medium approach (EMA) [4].

Theoretical model

In steel-making applications, it is assumed to manufacture pellets from the source powders, see Fig. 1. It is possible to fabricate pellets that are optimized for microwave absorption and reduction of iron ore if we know dependences of effective permittivity ε_{eff} and permeability μ_{eff} on the volume fraction of iron ore in the mixture.

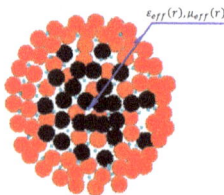

Figure 1. Model of pellet consisting of three types of source particles. Red particles are iron ore with permittivity ε_m and volume fraction of p_m; dark particles – biochar, ε_c, p_c; blue ones – binder, ε_b, p_b; white space between the particles is gas – ε_g, p_g.

The Bruggeman equation for three types of particles reads

Published under license by Materials Research Forum LLC.

$$(1 - p_m - p_c - p_b)\frac{\varepsilon_g - \varepsilon_{eff}}{\varepsilon_g + 2\varepsilon_{eff}} + p_m\frac{\varepsilon_m - \varepsilon_{eff}}{\varepsilon_m + 2\varepsilon_{eff}} + p_c\frac{\varepsilon_c - \varepsilon_{eff}}{\varepsilon_c + 2\varepsilon_{eff}} + p_b\frac{\varepsilon_b - \varepsilon_{eff}}{\varepsilon_b + 2\varepsilon_{eff}} = 0. \tag{1}$$

The generalized form of Bruggeman equation for N types of particles can be written as

$$\sum_{i=1}^{N} p_i \frac{\varepsilon_i - \varepsilon_{eff}}{\varepsilon_i + 2\varepsilon_{eff}} = 0. \tag{2}$$

In addition, the powders mixture under investigation can be considered as a mixture of two types of core-shell particles considering each particle of iron ore and biochar to be covered by binder (Fig. 2).

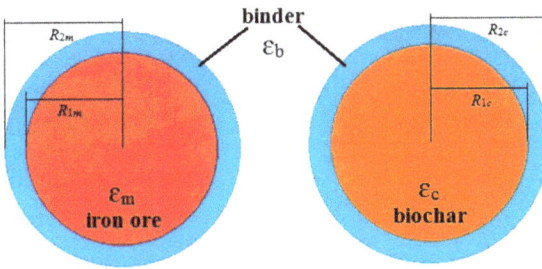

Figure 2. Model of two types of core-shell particles. Red particle are iron ore with permittivity ε_m and volume fraction of p_m; brawn particle – biochar, ε_c, p_c; blue shells – binder, ε_b, p_b; white space between the particles is gas – ε_g, p_g.

In this case, the effective medium equation according to EMA looks like following

$$(1 - p_m\zeta_m - p_c\zeta_c)\frac{\varepsilon_g - \varepsilon_{eff}}{\varepsilon_g + 2\varepsilon_{eff}} + p_m\zeta_m\frac{\varepsilon_b[3\varepsilon_m + (\zeta_m - 1)(\varepsilon_m + 2\varepsilon_b)] - \varepsilon_{eff}[3\varepsilon_b + (\zeta_m - 1)(\varepsilon_m + 2\varepsilon_b)]}{2\alpha_m\varepsilon_{eff} + \beta_m\varepsilon_b} +$$

$$+p_c\zeta_c\frac{\varepsilon_b[3\varepsilon_c + (\zeta_c - 1)(\varepsilon_c + 2\varepsilon_b)] - \varepsilon_{eff}[3\varepsilon_b + (\zeta_c - 1)(\varepsilon_c + 2\varepsilon_b)]}{2\alpha_c\varepsilon_{eff} + \beta_c\varepsilon_b} - p_m\zeta_m\frac{\frac{9}{2}\varepsilon_b(\varepsilon_m - \varepsilon_b)\ln(1 + l_m)}{2\alpha_m\varepsilon_{eff} + \beta_m\varepsilon_b} -$$

$$-p_c\zeta_c\frac{\frac{9}{2}\varepsilon_b(\varepsilon_c - \varepsilon_b)\ln(1 + l_c)}{2\alpha_c\varepsilon_{eff} + \beta_c\varepsilon_b} = 0. \tag{3}$$

Generalizing equation (3), the effective medium equation for N types of core-shell particles takes the form

$$(1 - \sum_{i=1}^{N} p_i\zeta_i)\left(\varepsilon_g - \varepsilon_{eff}\right)\prod_{i=1}^{N}(2\alpha_i\varepsilon_{eff} + \beta_i\varepsilon_{shell}) +$$

$$\left(\varepsilon_g - -2\varepsilon_{eff}\right)\sum_{i=1}^{N}\left[p_i\zeta_i\begin{Bmatrix}(\zeta_i - 1)(\varepsilon_i + 2\varepsilon_{shell})\left(\varepsilon_{shell} - \varepsilon_{eff}\right) + \\ +3\varepsilon_{shell}\left(\varepsilon_i - \varepsilon_{eff}\right) \end{Bmatrix} \times \\ \times \prod_{j=1,j\neq i}^{N}(2\alpha_j\varepsilon_{eff} + \beta_j\varepsilon_{shell})\right] - \left(\varepsilon_g - \right.$$

$$\left. -2\varepsilon_{eff}\right)\sum_{i=1}^{N}\left[\frac{9}{2}p_i\zeta_i\varepsilon_{shell}(\varepsilon_i - \varepsilon_{shell})\ln(1 + l_i) \times \\ \times \prod_{j=1,j\neq i}^{N}(2\alpha_j\varepsilon_{eff} + \beta_j\varepsilon_{shell})\right] = 0. \tag{4}$$

Shape Memory Alloys – SMA 2018   Materials Research Forum LLC
Materials Research Proceedings **9** (2018) 136-139   doi: http://dx.doi.org/10.21741/9781644900017-25

Results

According to [5,6] and the fact that $\varepsilon = \varepsilon' + \frac{4\pi\sigma}{\omega}i$ and $\omega = 2\pi v$ where v is a frequency of microwave radiation and equals to 2.45 GHz: $\varepsilon_m = 14.2 + 0.2i$, $\mu_m = 1.8 + 0.9i$, $\varepsilon_c = 3 + \frac{4\pi\sigma_c}{\omega}i$, $\mu_c = 1$, $\varepsilon_b = 1.2$, $\mu_b = 1$, $\varepsilon_g = 1$, $\mu_g = 1$, $p_b = 0.023$, $p_c = 0.9 - p_b - p_m$.

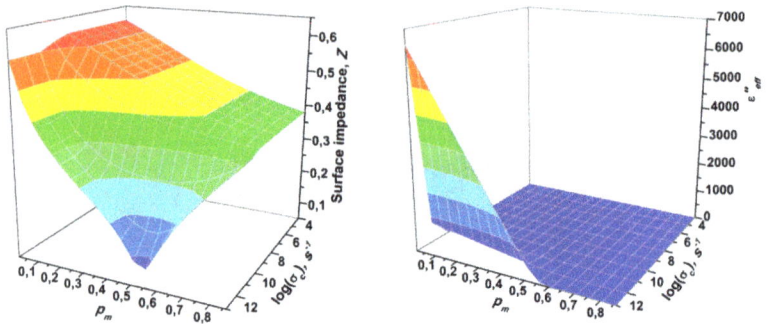

Figure 3. Surface impedance $Z = \sqrt{\frac{\mu_{eff}}{\varepsilon_{eff}}}$ and imaginary permittivity ε''_{eff} dependences on volume fraction of iron ore p_m and conductivity of biochar σ_c for the model of three types of spherical particles without shell.

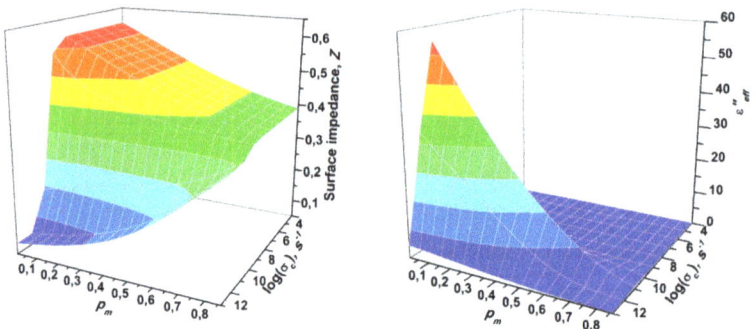

Figure 4. Surface impedance $Z = \sqrt{\frac{\mu_{eff}}{\varepsilon_{eff}}}$ and imaginary permittivity ε''_{eff} dependences on volume fraction of iron ore p_m and conductivity of biochar σ_c for the model of two types of core-shell particles.

Summary

In this paper, two approaches are considered for modeling the effective electrodynamic parameters of granules consisting of a mixture of iron ore, biochar, and bentonite as a binder powders. The first approach takes into account the fact that the binder envelopes the surface of each of the particles, therefore a model of particles with a shell is considered, where the material of the binder is used as the shell, and iron ore or biochar is used as the core. Thus, the first

Shape Memory Alloys – SMA 2018   Materials Research Forum LLC
Materials Research Proceedings **9** (2018) 136-139   doi: http://dx.doi.org/10.21741/9781644900017-25

approach deals with two types of core-shell particles. The second approach seems simpler, because considers three types of spherical particles without shell - iron ore, biochar, and bentonite binder.

However, it is shown on Fig. 3 and Fig. 4 that both approaches give an identical result when there are no conductive components, so it is advisable to use a simpler model of three types of particles without shell. Nevertheless, the use of a model of core-shell particles is justified in the case where the shell or core consists of a conducting material and it is necessary to take into account the effect of strong localization of the field near the skin layer on effective permittivity and permeability.

In addition, it is shown that the conductivity of biochar has a significant effect on the results for the problem of optimizing the penetration of microwave radiation into the examined granules. On the basis of dependencies Fig. 3 and Fig. 4, and assuming that wave impedance of surrounding environment is equal to 1, we can conclude that for better impedance matching there should be less iron ore concentration near the surface of pellet than in the core.

Acknowledgements

This work was supported by the Russian Foundation for Basic Research (Project No. 18-58-53055, 16-29-14045, 17-02-01382) and by the Ministry of Education and Science of the Russian Federation (State Contract No. 3.5698.2017/9.10).

References

[1] Zhiwei Peng, Jiann-Yang Hwang, Matthew Andriese, Absorber Impedance Matching in Microwave Heating, Appl. Phys. Express 5 (2012) 077301, 1-3.

[2] H. Wang, Z. Tian, L. Jiang, W. Luo, Wei Z., S. Li, J. Cui, W. Wei, Highly efficient adsorption of Cr(VI) from aqueous solution by Fe^{3+} impregnated biochar, J. Disper. Sci. Technol. 38 (2017) 815-825. https://doi.org/10.1080/01932691.2016.1203333

[3] J. Lee, K.-H. Kim, E.E. Kwon, Biochar as a Catalyst, Renew. Sust. Energ. Rev. 77 (2017) 70-79. https://doi.org/10.1016/j.rser.2017.04.002

[4] V.D. Buchelnikov, D.V. Louzguine-Luzgin, A.P. Anzulevich, G. Xie, S. Li, N. Yoshikawa, M. Sato, I.V. Bychkov, A. Inoue, Heating of metallic powders by microwaves: Experiment and theory, J. Appl. Phys. 1, 104 (2008) 1-10.

[5] M. P., Dielectric constant and magnetic permeability of various ferrites at ultrahigh frequencies, Sov. Phys. Usp. (Usp. Fiz. Nauk) 50 (1953) 152-155.

[6] R.S. Gabhi, D.W. Kirk, C.Q. Jia, Preliminary Investigation of Electrical Conductivity of Monolithic Biochar, Carbon, (2017) 11678. https://doi.org/10.1016/j.carbon.2017.01.069

Shape Memory Alloys – SMA 2018
Materials Research Proceedings 9 (2018) 140-143

Materials Research Forum LLC
doi: http://dx.doi.org/10.21741/9781644900017-26

The Influence of Cr_2O_3 and NiO on the Phase Transformation of Anatase-Rutile of Titanium Dioxide

Elena A. Belaya[1,a*], Valery V. Viktorov[2,b], Evgeny A. Belenkov[1,c]

[1]Chelyabinsk State University, 129 Brat'ev Kashirinykh Str., Chelyabinsk 454001, Russia

[2]South Ural State Humanitarian-Pedagogical University, 69 Lenin Prospect, Chelyabinsk 454091, Russia

[a]wea.csu@gmail.com, [b]viktorovvv@cspu.ru, [c]belenkov@csu.ru

*corresponding author

Keywords: Phase Transformations, Solid Solutions, Titanium Dioxide, Anatase, Rutile, Chromium Oxide, Nickel Oxide, Thermal Analysis, X-Ray Powder Diffraction

Abstract. In this paper, the phase transition of anatase to rutile by X-ray powder diffractometry and thermal analysis was studied, and the influence of chromium oxide and nickel oxide additives on this phase transformation. The anatase-rutile conversion was investigated by heating titanium dioxide in the temperature range from 20 °C to 1100 °C. As a result, it was established that the chromium and nickel oxides additions lower the onset temperature of the phase transformation and accelerate it. The observed regularities can be explained with the accepting the assumption of the martensitic mechanism of the anatase-rutile phase transformation.

Introduction

The temperature and rate of the anatase-rutile (A→R) phase transformation significantly depends on the synthesis conditions, the initial components purity, the accompanying impurities or specially introduced additives [1]. Additives can accelerate the phase transition and lower the temperature at which it occurs, or on the contrary, slow down it. Metals or their oxides are the additives that affect the A→R phase transition [2]. Thus, the anatase-rutile transformation is accelerated with the content of Nb, Cr, Si, and Fe in TiO_2. It was found that 1.0 mol.% of NiO, CoO, MnO_2, Fe_2O_3 or CuO promotes the conversion of anatase to rutile and the growth of TiO_2 grains. The phase transformation of anatase to rutile, depending on the chromium content (> 10 mmol.%), iron, silicon and especially the mixture of niobium and chromium, is also accelerated [3]. From a practical point of view, the additives introduction is necessary to adjust the ratio of anatase to rutile in synthesized titanium dioxide or to improve its photocatalytic characteristics [3]. The mechanism of the additives effect on the anatase-rutile phase transition is still not explained. Therefore, the aim of this work was to study the anatase-rutile phase transition in the presence of chromium and nickel oxides, as well as to study the effect of the initial oxides of TiO_2, NiO and Cr_2O_3 obtaining technology on this phase transition.

Experimental part

As the initial precursors, TiO_2, Cr_2O_3, $NiCO_3$ and $NiNO_3 \cdot 6H_2O$ (analytical grade) were taken. The initial titanium (IV) oxide in anatase modification was obtained in two ways:

Synthesis 1 – Hydrolytic titanium dioxide (HTO) was obtained by thermal hydrolysis of Ti (IV) sulfate compounds solutions, which was annealed at 600 °C for 2 hours.

Synthesis 2 – Thermal hydrolysis of $TiCl_4$ detailed described in [4].

By thermolysis of $NiCO_3$ and $NiNO_3 \cdot 6H_2O$ at a temperature of 800 °C for 3 hours nickel oxide was obtained. The purity of the initial substances was controlled on the spectrometer PGS 2 and X-ray spectral analyzer CPM 25. The initial mechanical mixtures of the systems 95 wt.% TiO_2 – 5 wt.% Cr_2O_3 and 95 wt.% TiO_2 – 5 wt.% NiO were prepared from the

Published under license by Materials Research Forum LLC.

Shape Memory Alloys – SMA 2018   Materials Research Forum LLC
Materials Research Proceedings **9** (2018) 140-143   doi: http://dx.doi.org/10.21741/9781644900017-26

previously synthesized titanium (IV) oxide and chromium (III) oxide or nickel oxide by thoroughly mixing in an agate mortar up to obtaining uniform color powders, after which it was calcined in the temperature range 600–1200 °C for 1 hour.

For the TiO_2-NiO system, four mixtures were prepared:

Mixture I: 95 wt.% TiO_2 (synthesis 1) – 5 wt.% NiO, obtained from $Ni(NO_3)_2$;

Mixture II: 95 wt.% TiO_2 (synthesis 1) – 5 wt.% NiO (synthesized from basic carbonate);

Mixture III: 95 wt.% TiO_2 (synthesis 2) – 5 wt.% NiO, obtained from $Ni(NO_3)_2$;

Mixture IV: 95 wt.% TiO_2 (synthesis 2) – 5 wt.% NiO (synthesized from basic carbonate).

Chemical analysis of the initial mixtures using the X-ray spectral analyzer CPM 25 and the spectrometer PGS 2 was carried out using standard techniques. Quantitative and qualitative phase analysis of the samples was carried out on a diffractometer Rigaku Ultima IV with Kα-radiation of copper. Thermal analysis (TG-DSC) was carried out on a thermal analyzer Netzch 449C "Jupiter" at a heating rate of 10 °C/min to 1100 °C in corundum crucibles in argon flow.

Results and discussion

On powder X-ray diffraction patterns of the initial titanium dioxide synthesized by methods 1 and 2, only diffraction maxima characteristic of anatase were observed. Synthesized anatase was heated.

Thermoanalytical curves for anatase obtained from HTO (synthesis 1) are shown in Fig. 1. At the initial stage, up to a temperature of 200 °C, a small loss of mass and the corresponding endoeffect associated with the removal of water are fixed.

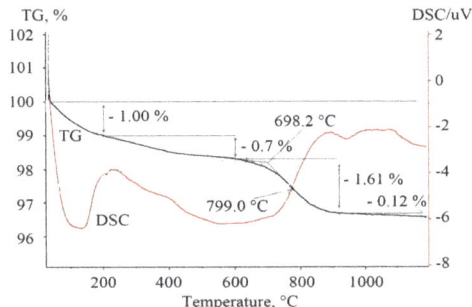

Figure 1. Thermal analysis results of the anatase sample synthesized by the first method.

At a temperature of the order of 700–710 °C, an active desulfatization of the samples begins. Exoeffect at a temperature of ≈ 850 °C can be connected with the phase transformation A → R. For the sample obtained by synthesis 2, thermal analysis shows a single exoeffect in the region of ≈ 900 °C, which can be attributed to the beginning of the polymorphic transformation A → R.

The conducted X-ray phase analysis confirms the results of thermal analysis, according to which for the anatase synthesized by the first method, the characteristic peaks corresponding to the rutile modification are fixed at 850 °C, and at 950 °C the anatase phase completely disappears. While for the titanium dioxide sample obtained by the second method, the rutile modification appears at 950 °C, and the formation of the rutile pure phase occurs 100 °C higher.

For mixtures of TiO_2-Cr_2O_3 and TiO_2-NiO similar investigations were carried out.

On the thermoanalytical curves of a sample containing 5 wt.% and 95 wt % of TiO_2 (Fig. 2a), the exo-effect corresponding to the anatase-rutile phase transition occurs over a wide temperature range, and the rutile nuclei are formed at lower temperatures than in pure anatase.

For the mixtures I-IV system TiO_2-NiO (Fig. 2b) the thermograms are almost identical to the anatase thermogram, only endo-effects associated with the decomposition of nitrate or basic nickel carbonate are present.

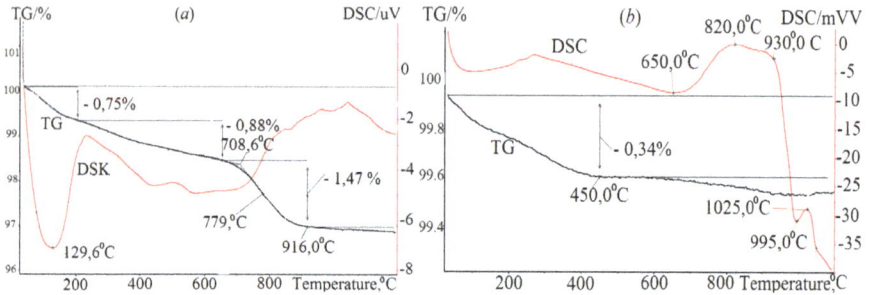

Figure 2. Thermal analysis results: (a) mixtures 95 wt.% TiO_2 – 5 wt.% NiO (mixture II), (b) mixtures of 95 wt.% TiO_2 – 5 wt.% Cr_2O_3.

X-ray diffraction analysis revealed that in the initial stages of the process, with and without addition of NiO or Cr_2O_3, processes of the anatase crystals formation and their growth take place. Intensive dissolution of chromium oxide, with the formation of a solid solution and the nickel titanate formation, begins only with the rutile phase appear [5].

Both chromium oxide and nickel oxide are additives that stimulate the anatase phase transition into rutile, substantially shifting the time and temperature interval of the phase transition (Fig. 3a, b).

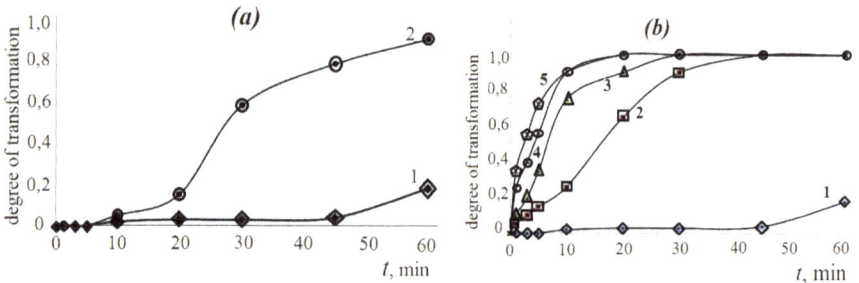

Figure 3. The conversion degree of anatase to rutile at isothermal exposure 850 °C: (a) – curve 1 – TiO_2 without addition of Cr_2O_3; curve 2 – TiO_2 with the addition of Cr_2O_3; (b) curve 2 – mixture II; curve 3 – mixture I; curve 4 – mixture of IV; curve 5 – mixture III.

The appearance of the rutile phase in the pure anatase obtained from the synthesis 1, as shown on Fig. 3(a) and 3(b), is observed on 40 minutes of calcination at 850 °C, and with the addition of Cr_2O_3, the formation of the rutile phase was observed 20 minutes earlier at the same temperature.

Nickel oxide affected the A → R phase transition more efficiently. Thus, the rutile phase formation at 850 °C (Fig. 3b) was observed 40 minutes earlier. In addition, in the case of NiO additives, the temperature and time shift of the anatase-rutile phase transformation substantially depends on the initial way of nickel oxide synthesis. So for NiO, obtained from nickel nitrate, the rutile phase appears earlier than for NiO, synthesized from the basic carbonate [6]. The most effective reduction is observed for mixture III.

The reason of the observed phenomena can be a martensitic rather than diffusion mechanism of the polymorphic transformation A → R. The rate of diffusion with the same chemical composition of the system is determined primarily by the temperature, so a decrease in the polymorphic transformation temperature or an increase in the rate of its flow at the identical temperatures cannot be explained by the diffusion mechanism. Evidently, the anatase restructuring into rutile structure occurs through a martensite mechanism as a result of the anatase crystal lattice transformation so that the atoms displacements, leading to a rearrangement of its structure, occur collectively, on distances not exceeding interatomic distances. Such processes occur with the right proportionally of the rutile and anatase crystal lattices at the interphase boundary, where the phase transformation proceeds. The stimulating effect of impurities on the phase transformation can be explained by the fact that they contribute to the proportionality of the lattices and provide a martensitic rather than diffusional character of the A → R phase transition.

Summary

The influence of NiO and Cr_2O_3 on the anatase-rutile phase transition was investigated by X-ray diffraction and thermal (TG-DSC) analyzes. Nickel and chromium oxides are additives that reduce the temperature of the anatase phase transition into rutile and accelerate the phase transition. Nickel oxide efficiently reduces the initial temperature of the phase transformation and decreases the rutile formation time. The temperature of the polymorphic transformation beginning depends on the initial way of the preparation of both nickel (II) oxide and titanium (IV) oxide. In the temperature range of 700–850 °C nickel titanate $NiTiO_3$ is formed only from the anatase, while the anatase completely passes into rutile. A solid solution of Cr_2O_3 in TiO_2 is formed only on the rutile modification basis. The stimulating effect of additives on the phase transformation can be explained by the fact that they contribute to the proportionality of the lattices and provide a martensitic rather than diffusional character of the A → R phase transition.

Reference

[1] V.V. Viktorov, E.A. Belaya, A.S. Serikov, Phase Transformations in the System TiO_2–NiO, Inorg. Mat. 48 (2012) 570-575. https://doi.org/10.1134/S0020168512050202

[2] F. Matteucci, G. Cruciani, M. Dondi, M. Raimondo, The Role of Counterions (Mo, Nb, Sb, W) in Cr-, Mn-, Ni- and V-doped Rutile Ceramic Pigments. Part 1. Crystal structure and phase transformations, Ceram. Int. 32 (2006) 385-392. https://doi.org/10.1016/j.ceramint.2005.03.014

[3] S. Karvinen, The Effects of Trace Elements on the Crystal Properties of TiO_2, Solid State Sci. 5 (2003) 811-819. https://doi.org/10.1016/S1293-2558(03)00082-7

[4] E.E. Movsesov, L.P. Sedova, T.N. Grishina, Patent RF, 2049066 (1995).

[5] E.A. Belaya, V.V. Viktorov, Formation of Solid Solutions in the TiO_2-Cr_2O_3 System, Inorg. Mat. 44 (2008) 62-66. https://doi.org/10.1134/S002016850801010X

[6] V.V. Viktorov, A.S. Serikov, E.A. Belaya, Phase Transformations in the TiO_2–NiO System, Inorg. Mat., 48 (2012) 488-493. https://doi.org/10.1134/S0020168512050202

Shape Memory Alloys – SMA 2018   Materials Research Forum LLC
Materials Research Proceedings **9** (2018) 144-147   doi: http://dx.doi.org/10.21741/9781644900017-27

Martensitic Transformations of Carbon Polytypes

Evgeny A. Belenkov[1,a*], Vladimir A. Greshnyakov[1,b]

Chelyabinsk State University, 129 Brat'ev Kashirinykh Str., Chelyabinsk 454001, Russia

[a]belenkov@csu.ru, [b]greshnyakov@csu.ru

*corresponding author

Keywords: Phase Transition, Computer Simulation, Polytypism, Graphite, Diamond

Abstract. In this paper, the mutual phase transformations of graphite and diamond polytypes were investigated by the density functional theory methods and the atom-atom potential. It was found out that a potential barrier with a height of 6.3 J/mol must be overcome to shift the graphene layers in graphite, and for the shift of the molecular layer in diamond, the potential barrier is 216 kJ/mol. Mutual structural transformations of almost any polytypes are possible due to random shifts of separate layers. Shift transformations can be the reason for the formation of crystals with random packing of layers and the impossibility of forming ideal polytypes.

Introduction

Structural varieties of carbon compounds can be divided into allotropes, polymorphs and polytypes [1-6]. The polytypes differ from each other in the packing order of molecular layers [3,4]. The most known polytypes are 2H and 3R graphite polytypes, as well as 2H and 3C diamond polytypes. For each of the polymorphic varieties of carbon, the existence of hundreds of different polytypes differing in the order of molecular layer packing is possible [7].

Mutual phase transformations of some allotropes, polymorphs and polytypes into others are possible. In phase transformations of allotropes, the atomic coordination in the structure changes. For example, during the phase transition of graphite to diamond, three-coordinated atoms become four-coordinated atoms, which corresponds to the transition of atoms from sp^2-hybridized states to sp^3-hybridized states [4,5]. In the case of mutual transformations of the polymorph structures, the change in the relative arrangement of atoms in the first coordination sphere should occur, as a result of which the bond lengths and bond angles change [6]. Phase transitions of some polytypes into others should occur as a result of changes in the packaging order of layers. The mechanism of these transformations is insufficiently studied, therefore the mutual transformations of carbon polytypes are theoretically investigated in this work.

Calculation methods

As an initial assumption, it was accepted that the transformation of the polytype structure can occur as a result of collective shifts in molecular layers. Such structural transformations can be considered as an analog of martensitic transitions. Calculations of structural transformations of polytypes were performed for graphite and diamond crystals.

Simulation of relative shifts of graphene layers in graphite crystals was performed by the atom-atom potential method [8]. In diamond crystals, similar calculations of molecular layer shifts were performed using the density functional theory method in the local density approximation [9] and the generalized gradient approximation [10]. As a result, the magnitudes of the potential barriers that need to be overcome to shift the layers in graphite and diamond have been found.

Published under license by Materials Research Forum LLC.

Shape Memory Alloys – SMA 2018   Materials Research Forum LLC
Materials Research Proceedings **9** (2018) 144-147   doi: http://dx.doi.org/10.21741/9781644900017-27

Results and discussion

At the next stage, all possible mutual transformations of polytypes containing in the unit cell from two to thirteen molecular layers were considered. Theoretical analysis has shown that the shift of only hexagonal layers can occur. For example, the "ABA" fragment of the polytype structure can be transformed into the «ACA» structure by shifting the layer from the position "B" to "C". In Fig. 1 an example of the mutual structure transformation of 2H polytype into 4H polytype, 2H polytype into 6H(1) polytype and 2H polytype into 6H(2) polytype, resulting from successive shifts of molecular layers, is given. The modeling of structural transformations as a result of random shifts of the layers showed that the only polytype that is not subject to structural change is 3C diamond polytype. All other polytypes can experience constant mutual transformations so that ideal polytype structures can be observed very rarely during this process. On average, with such transformations, the number of hexagonal and nonhexagonal layers in the structures under consideration should be approximately the same.

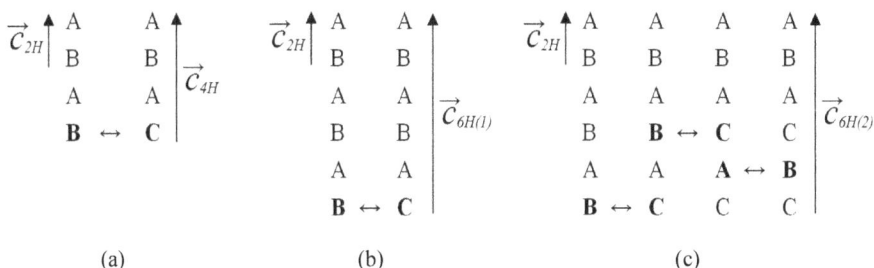

Figure 1. Schemes of mutual transformation of the polytype structures: (a) 2H ↔ 4H; (b) 2H ↔ 6H(1); (c) 2H ↔ 6H(2).

In the simulation of polytype phase transitions, it was assumed that they could occur due to shifts of graphene layers in graphite crystals or molecular layers in diamond crystals. Calculations of the activation energy of the shifts were performed by the atomic-atomic potential method for graphite. The results of calculations of the binding energy of neighboring layers as a function of the shift are shown in Fig. 2. The height of the potential barrier, which must be overcome to shift the layer from one local minimum to the other, is 6.25 J/mol. Therefore, the ordered structure can be absent at room temperature in graphite nanocrystals, and structural transformations of some polytypes into others can occur continuously.

The activation energy of the diamond layer was calculated by the density functional theory method for the successive shift of two atoms (with numbers 4 and 5 in Fig. 3) in the XY plane from the position $(2a/3; b/3)$ to the position $(a/3; 2b/3)$. In diamond, the molecular layer shift energy determined from the graph in Fig. 4 is 213-216 kJ/mol. Such shifts of layers can cause phase transformations of the structure of any noncubic polytype into the structure of any other polytype. In addition, shear transformations can be the cause of the formation of crystals with random packing of layers [6]. Phase transitions of some polytypes into others are possible in nanodiamonds at high temperatures, whereas in diamond macrocrystals these transitions may be absent due to the significant value of the activation energy necessary for the shift of the molecular layer.

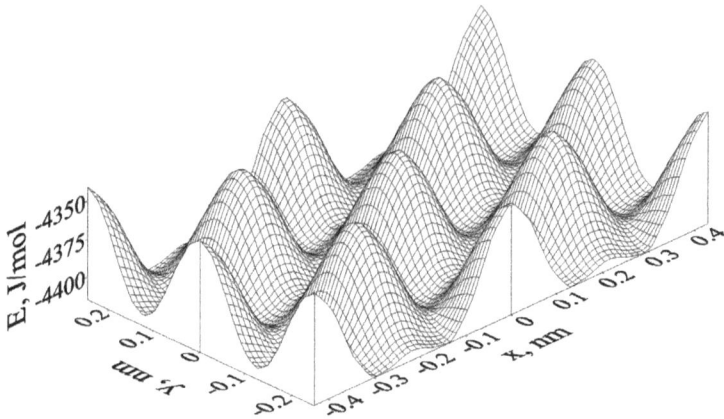

Figure 2. Dependence of the van der Waals energy (E) on the shifts of the graphene layer along the X and Y axes.

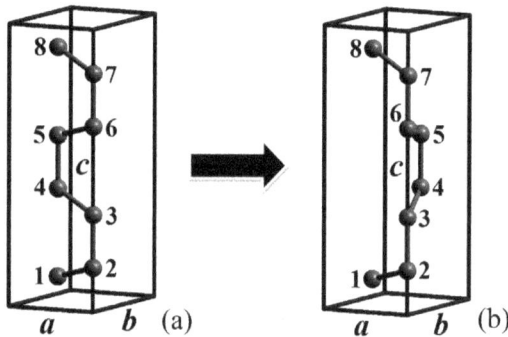

Figure 3. Transformation of the 2H diamond polytype unit cell into the 4H diamond polytype unit cell (atoms 4 and 5 are shifted).

Summary

Thus, theoretical calculations of the possibility of martensite transitions in carbon polytypes are performed in this paper. The polytype structures differ only in the packing order of the molecular layers, and their mutual phase transformations are possible with shifts of these layers. Calculations using the atom-atom potential method have shown that a potential barrier of 6.3 J/mol must be overcome to shift the graphene layers in graphite. The height of the similar barrier in diamond crystals found by the DFT method was equal to 213–216 kJ/mol. Successive shifts of individual layers in graphite or diamond crystals can lead to a transformation of the structure of some polytypes into others. The only exception is 3C diamond polytype, the ideal structure of which cannot be transformed. If random shifts of the layers in the crystals are possible, then the layer packing will be random. Therefore, crystals with the structure of ideal polytypes can not be formed.

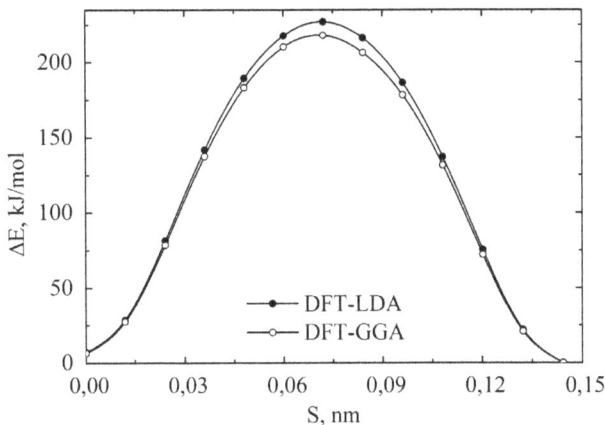

Figure 4. Dependences of the difference total energy (ΔE) on the modulus of the shear vector (S) by the molecular layer shifts when 2H polytype transforms into a 4H polytype.

References

[1] E.A. Belenkov, V.A. Greshnyakov, Classification of structural modifications of carbon, Phys. Solid State 55 (2013) 1754-1764. https://doi.org/10.1134/S1063783413080039

[2] E.A. Belenkov, V.A. Greshnyakov, Classification schemes for carbon phases and nanostructures, New Carbon Mater. 28 (2013) 273-283. https://doi.org/10.1016/S1872-5805(13)60081-5

[3] E.A. Belenkov, V.A. Greshnyakov, Structural varieties of polytypes, Phys. Solid State 59 (2017) 1926-1933. https://doi.org/10.1134/S1063783417100055

[4] V.A. Greshnyakov, E.A. Belenkov, Investigation on the formation of lonsdaleite from graphite, J. Exp. Theor. Phys. 124 (2017) 265-274. https://doi.org/10.1134/S1063776117010125

[5] E.A. Belenkov, V.A. Greshnyakov, Diamond-like phases prepared from graphene layers, Phys. Solid State 57 (2015) 205-212. https://doi.org/10.1134/S1063783415010047

[6] E.A. Belenkov, V.A. Greshnyakov, Structure, properties, and possible mechanisms of formation of diamond-like phases, Phys. Solid State 58 (2016) 2145-2154. https://doi.org/10.1134/S1063783416100073

[7] Yu.E. Kitaev, M.I. Aroyo, J.M. Perez-Mato, Site symmetry approach to phase transitions in perovskite-related ferroelectric compounds, Phys. Rev. B 75 (2007) 064110. https://doi.org/10.1103/PhysRevB.75.064110

[8] A.I. Kitaigorodsky, Molecular crystals and molecules, Academic Press, New York, 1973.

[9] J.P. Perdew, A. Zunger, Self-interaction correction to density-functional approximations for many-electron systems, Phys. Rev. B 23 (1981) 5048-5079. https://doi.org/10.1103/PhysRevB.23.5048

[10] J.P. Perdew, K. Burke, M. Ernzerhof, Generalized gradient approximation made simple, Phys. Rev. Lett. 77 (1996) 3865-3868. https://doi.org/10.1103/PhysRevLett.77.3865

Shape Memory Alloys – SMA 2018
Materials Research Proceedings 9 (2018) 148-151

Materials Research Forum LLC
doi: http://dx.doi.org/10.21741/9781644900017-28

Martensitic Structural Transformations of Fluorographene Polymorphic Varieties

Maxim E. Belenkov[1,a]*, Vladimir M. Chernov[1,b], Evgeny A. Belenkov[1,c]

[1]Chelyabinsk State University, 129 Brat'ev Kashirinykh Str., Chelyabinsk 454001, Russia

[a]me.belenkov@gmail.com, [b]chernov@csu.ru, [c]belenkov@csu.ru

*corresponding author

Keywords: Graphene, Fluorographene, Polymorphism, Phase Transitions, Modeling

Abstract. The structure and energy characteristics of the five main structural varieties of fluorographene were calculated by the methods of the density functional theory in the generalised gradient approximation. It was established that the structural type of T1 has maximum sublimation energy. Other polymorphic varieties of fluorographene T2–T5 can be converted into a T1 structure under uniaxial stretching of layers. Such structural transformations of fluorographene layers are similar to martensitic phase transitions.

Introduction

The functionalization of graphene for practical applications is possible as a result of chemical adsorption of hydrogen, fluorine, oxygen or chlorine atoms on its surface [1-4]. The existence of five basic structural types of functionalized graphene layers is possible [5,6]. The graphene modified by fluorine has a number of advantages over graphene layers modified by hydrogen, oxygen or chlorine due to its greater stability and lower toxicity. It is possible to obtain structural varieties as a result of their mutual phase transformations. In this paper, polymorphic varieties of fluorographene and possible ways of their mutual structural transformations were studied theoretically.

Calculation Methods

Based on the theoretical analysis it was established that only five methods of attaching fluorine atoms to hexagonal graphene layers are possible, where structural positions of all carbon atoms are equivalent [6]. These polymorphic varieties of fluorographene should be the most thermodynamically stable, therefore they were chosen as the object for research.

The fluorographene layers of the five main types T1–T5, modeled by the addition fluorine to hexagonal graphene layers, were geometrically optimized by the methods of the density functional theory (DFT) in the generalized gradient approximation (GGA). Calculations of the structure and total energy of the layers were performed using the Quantum ESPRESSO software package. The calculations were performed for bulk structures, namely, for stacks of fluorographene layers with an interplanar distance of 10 Å. This distance ensured the absence of the influence of the adjacent layers on each other, so the calculated structure of the layers and their properties corresponded to the values characteristic for isolated layers. The density of electronic states was calculated using a k-point grid: $12 \times 12 \times 12$. The sublimation energy (E_{sub}) of fluorographene polymorphs was calculated as the difference between the total specific energy attributable to the molecular group CF and the energy of isolated fluorine and carbon atoms.

In the modeling of phase transitions, polymorphic transformations of T2–T5 layers into layers of the first type T1 were examined. For this, extended unit cells of the layers were chosen, so that the cell of one layer could be transformed into the unit cell of the other using stretching. The selected elementary cells of the layers must contain the same number of atoms and belong to the same crystal system. The minimal hexagonal unit cell of the T1 layer contains only 4 atoms (Table 1). Therefore, for the T1 layer, extended orthorhombic unit cells were considered. For T1,

Published under license by Materials Research Forum LLC.

Shape Memory Alloys – SMA 2018   Materials Research Forum LLC
Materials Research Proceedings **9** (2018) 148-151   doi: http://dx.doi.org/10.21741/9781644900017-28

T2, and T3 layers, 4 fluorine atoms and 4 hydrogen atoms were contained in the commensurate unit cells. For the T1, T4 and T5 layers, commensurate unit cells contained 8 CF molecules. Images of commensurate unit cells for layers T1 and T3 are shown in Fig. 1. The mutual phase transformations in this case occur as a result of stretching-compression along the direction specified by the vector b. Modeling of structural transformations was performed as a result of the calculation of successive states of fluorographene layers of different types undergoing uniaxial compression or stretching.

Results and Discussion
The results of calculations of the geometrically optimized structure of the main polymorphs using the DFT-GGA method are shown in Fig. 2, the numerical values of the structural parameters and energies are given in Table 1. Variations in the structure of polymorphic fluorographene species are caused by different order of addition of fluorine atoms to the graphene layer and because of this, the degree of corrugation differs significantly in the layers. The least corrugated CF layer is the T1 layer of the first structural type.

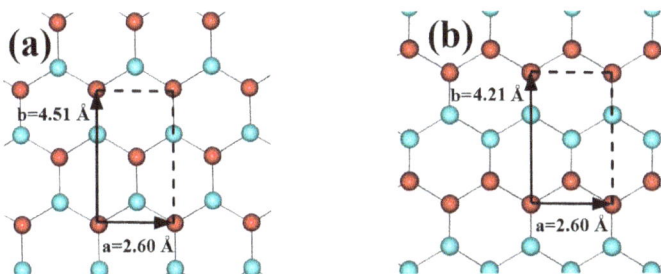

Figure 1. Unit cells of fluorographene layers: (a) T1; (b) T3 (different colors denote the carbon atoms where fluorine is attached from different sides of the layer).

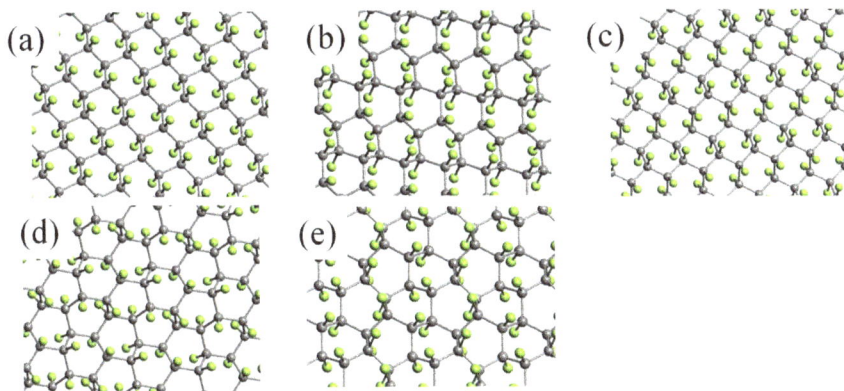

Figure 2. Geometrically optimized structure of polymorphic varieties of fluorographene layers: (a) T1; (b) T2; (c) T3; (d) T4; (e) T5.

The maximum energy of sublimation is observed for a fluorographene layer of the first structural type. This polymorphic type must have the greatest thermodynamic stability. Therefore, other polymorphs of fluorinated graphene should strive to transform their structures into the structure of T1. Polymorphic transformations are possible as a result of uniaxial stretching and compression of the layers. For example, for layers of the first three types, significant differences are observed in the length of the elementary translation vector b, so the T2 and T3 layers can be transformed into the T1 layer under uniaxial stretching along the direction specified by this vector. The results of the phase transformation calculations for the T1 and T3 layers are shown in Fig. 3. The phase transformation should occur when the length of the vector b becomes 4.32 Å. The height of the potential barrier in the transition T3 → T1 is 147 meV / u.c. With the reverse transformation T1 → T3 ΔE = 171 meV / u.c.

Table 1. The structural parameters and properties the main types of fluorographene layers (T1–T5), as well as hexagonal (L_6) graphene (unit cell parameters a and b, sublimation energy E_{sub}, numbers of atoms N in minimal unit cell).

Type of layer	T1	T2	T3	T4	T5	L_6
Crystal system	Hex	Ort	Ort	Ort	Ort	Hex
a [Å]	2.60	2.58	2.60	4.88	5.06	2.49
b [Å]		4.57	4.21	4.58	4.62	
γ [°]	120	90	90	90	90	120
E_{total} [eV/u.c.]	−1601.27	−3202.04	−3202.53	−6403.18	−6404.19	−314.64
E_{total} [eV/(CF)]	−800.64	−800.51	−800.63	−800.40	−800.52	−
E_{sub} [eV/(CF)]	14.32	14.19	14.31	14.08	14.20	−
N	4	8	8	16	16	2

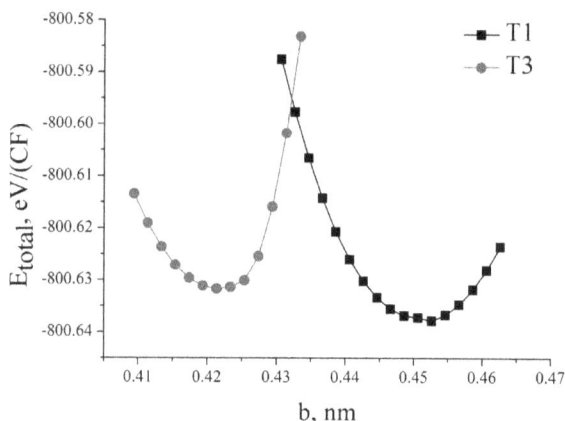

Figure 3. Graph of the dependence of the total energy E_{total} of the unit cells T1 and T3 of the fluorographene layers upon their compression and extension along the direction specified by the vector b.

During the phase transformation, the order of addition of fluorine atoms to the graphene layer should also change. This can occur due to a 180° rotation of the molecular group C_2F_2, as a result of which fluorine atoms attached to the layer from top and bottom are interchanged. Similar rotations, but at an angle of 90°, occur with C_2 molecular fragments in the graphene layers under stretching, which causes the formation of Stone-Wales defects in the layer and their movement occurs. Obviously, similar rotations are also possible in fluorographene layers. The transformation of the fluorographene layers structure should apparently occur in an avalanche manner, when the rotation of one molecular fragment in the layer gives a push to the same rotations of neighboring C_2F_2 fragments. Therefore, the phase transformation process that has begun should proceed at high rates. Polymorphic transformations of structural fluorographene varieties can be considered as an analog of martensitic phase transitions occurring in phases with a bulk structure. To estimate the activation energy of the rotation of the molecular group C_2F_2, three methods have been calculated for which: carbon-carbon bonds are broken, the molecule rotates 180° and the bonds form again; there is no complete break in the interatomic bonds, but the molecule rotates 180° around the axis specified by the carbon-carbon bond; in the latter case, only a simultaneous shift of the fluorine atoms occurs, so that adjacent carbon atoms exchange the fluorine atoms located on the different sides of the layer. The minimum energy of the potential barrier that must be overcome to transform the structure corresponds to the third method and is ~5.5 eV / u.c.

Summary

Thus, as a result of calculations by the DFT-GGA method, it is established that out of the five main polymorphic varieties of fluorographene, the maximum sublimation energy is observed for polymorphic type T1, which should be the most thermodynamically stable. Phase transformations of the remaining fluorographene polymorphs to the T1 structure possible under uniaxial compression-stretching of the layers. These structural transformations are similar to martensitic phase transitions.

References

[1] R.R. Nair, W. Ren, R. Jalil, I. Riaz, V.G. Kravets, L. Britnell, P. Blake, F. Schedin, A.S. Mayorov, S. Yuan, M.I. Katsnelson, H.M. Cheng, W. Strupinski, L.G. Bulusheva, A.V. Okotrub, I.V. Grigorieva, A.N. Grigorenko, K.S. Novoselov, A.K. Geim, Fluorographene: A two dimensional counterpart of Teflon, Small 6 (2010) 2877. https://doi.org/10.1002/smll.201001555

[2] D.C. Elias, R.R. Nair, T.M.G. Mohiuddin, S.V. Morozov, P. Blake, M.P. Halsall, A.C. Ferrari, D.W. Boukhvalov, M.I. Katsnelson, A.K. Geim, K.S. Novoselov, Control of graphene's properties by reversible hydrogenation: evidence for graphane, Science 323 (2009) 610-613. https://doi.org/10.1126/science.1167130

[3] D. Chen, H. Feng, J. Li, Graphene oxide: preparation, functionalization, and electrochemical applications, Chem. Rev. 112 (2012) 6027-6053. https://doi.org/10.1021/cr300115g

[4] B. Li, L. Zhou, D. Wu, W. Peng, K. Yan, Y. Zhou, Z. Liu Photochemical chlorination of graphene, ACS Nano 5 (2011) 5957-5961. https://doi.org/10.1021/nn201731t

[5] T.E. Belenkova, V.A. Greshnyakov, V.M. Chernov, E.A. Belenkov, Structure of graphane polymorphs, J. Phys. Conf. Ser. 917 (2017) 032015. https://doi.org/10.1088/1742-6596/917/3/032015

[6] E.A. Belenkov, V.M. Chernov, E.A. Belenkov, Structure of fluorographene and its polymorphous varieties, in: Book of abstracts «SPBOPEN 2018», Academic University Publishing, St. Petersburg, 2018, pp. 30-31.

Shape Memory Alloys – SMA 2018
Materials Research Proceedings 9 (2018) 152-156

Materials Research Forum LLC
doi: http://dx.doi.org/10.21741/9781644900017-29

Diamond-Like Phase Transformations of Martensitic Type

Vladimir A. Greshnyakov[1,a*], Evgeny A. Belenkov[1,b]

[1]Chelyabinsk State University, 129 Brat'ev Kashirinykh Str., Chelyabinsk 454001, Russia

[a]greshnyakov@csu.ru, [b]belenkov@csu.ru

*corresponding author

Keywords: Diamond, Diamond-Like Phases, Polymorphism, Phase Transitions, Modeling

Abstract. Mutual phase transformations of polymorphic varieties of diamond were theoretically investigated by the density functional theory methods. The local density approximation and the generalized gradient approximation were used in the calculations. As a result of the calculations, it was found that the 2H diamond polytype can be obtained by compressing the 3C diamond polytype in the [110] and [211] crystallographic directions at pressures from 300 to 380 GPa. The compression of the diamond-like LA3 phase in the [100] direction results in the transformation of its structure into the structure of LA5 phase (the phase transition pressure is 160–169 GPa). At these phase transformations, the atoms in the structures are displaced by distances smaller than the interatomic distances, therefore they can be regarded as martensitic transformations.

Introduction

Martensitic types transitions occur in pure metals and their alloys, as well as in crystals with ionic, covalent or van der Waals type of chemical bonds [1]. In martensitic transformations, the change in the mutual arrangement of the atoms forming the crystal occurs as a result of their ordered displacement so that the relative displacements of neighboring atoms are small in comparison with the interatomic distance [2]. The possibility of martensitic transformations in diamond-like materials is unexplored, since phase transformations of diamond-like phases occur at ultrahigh pressures, which makes their experimental studies more difficult. The possibility of the existence of a large variety of diamond-like phases is predicted theoretically [3,4]. Therefore, in this paper, theoretical studies of polymorphic transformations of diamond-like phases at high pressures were carried out.

Calculation methods

Modeling of the "cubic 3C diamond → 2H diamond polytype" and "LA3 phase → LA5 phase" phase transitions was performed by the density functional theory method for a number of intermediate structures in the transformation process of the structure of one phase into another using the method described in [5-7]. The calculations were carried out using the Perdew-Zunger (local density approximation) and Perdew-Burke-Ernzerhof (general gradient approximation) exchange-correlation energy functionals. The effect of ion cores was taken into account by using norm-conserving pseudopotentials. We used a $12 \times 12 \times 12$ grid of k-points in our calculations. Wave functions were decomposed in terms of a truncated basis set of plane waves. To limit the dimension of the set of basis functions, E_{cutoff} was assumed to be 60 Ry.

Results and discussion

To calculate the phase transformations, the elementary cells containing the same number of atoms equal to sixteen were chosen in the structures of the initial and final phases. Images of the unit cells of cubic and hexagonal diamonds, as well as diamond-like LA3 and LA5 phases are shown in Fig. 1.

Published under license by Materials Research Forum LLC.

Shape Memory Alloys – SMA 2018
Materials Research Proceedings **9** (2018) 152-156

Materials Research Forum LLC
doi: http://dx.doi.org/10.21741/9781644900017-29

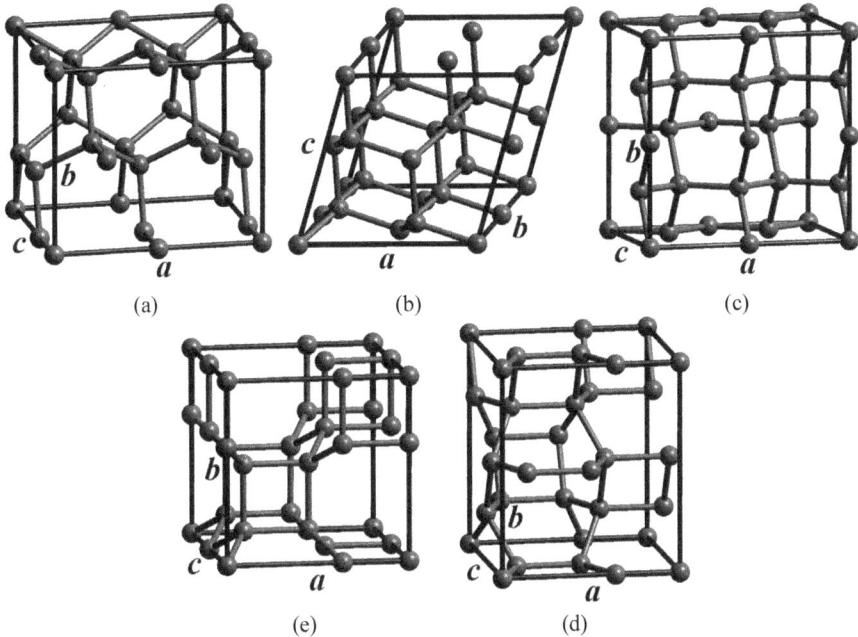

Figure 1. Unit cells of polymorphic varieties of diamond: (a) and (b) cubic 3C diamond;
(c) hexagonal 2H diamond; (d) LA3 phase; (e) LA5 phase.

The calculations showed that hexagonal diamond may be obtained by the uniaxial static compression of cubic diamond along the crystallographic [110]-axis. Fig. 2 represents the calculations of difference total energy (ΔE_{tot}) and volume (V) per atom for the cases of compression and decompression of the diamond-like compounds. It was established that hexagonal diamond strained in the [110]-direction can be obtained from cubic diamond at the volume decrease to 4.745 (4.983) Å^3/atom (Fig. 2) corresponding to the pressure of 380 (300) GPa. Pressures corresponding to different deformation extents were calculated as the total energy second derivative with respect to volume [5,6].

Another way of obtaining the 2H diamond is uniaxial compression of cubic diamond along the crystallographic [211]-axis (Fig. 3). In this case, hexagonal 2H diamond polytype, deformed along the [100]-axis, is created at $V \sim 5.0$ Å^3/atom corresponding to the pressure of 303 (354) GPa.

Thus, hexagonal diamond may be obtained by compressing of cubic diamond. Pressures necessary to form hexagonal diamond by compressing the 3C diamond along the [110]- and [211]-axes may be assumed to be approximately equal to each other and range from 300 to 380 GPa. These results agree well with experimental data from paper [8] devoted to studying pulsed laser irradiation of pyrolytic graphite resulting first in the graphite transformation into the 3C diamond at 60 GPa and then, at the pressure of ~215 GPa, in the 3C diamond transformation into the 2H diamond polytype.

At the final stage of the research it was established that orthorhombic diamond-like LA5 phase can be formed on the basis of tetragonal LA3 phase upon compression of its structure along the [100]-axis. In this case, the phase transition pressure is 160 (169) GPa with the atomic volume of 5.434 (5.576) Å³/atom (Fig. 4).

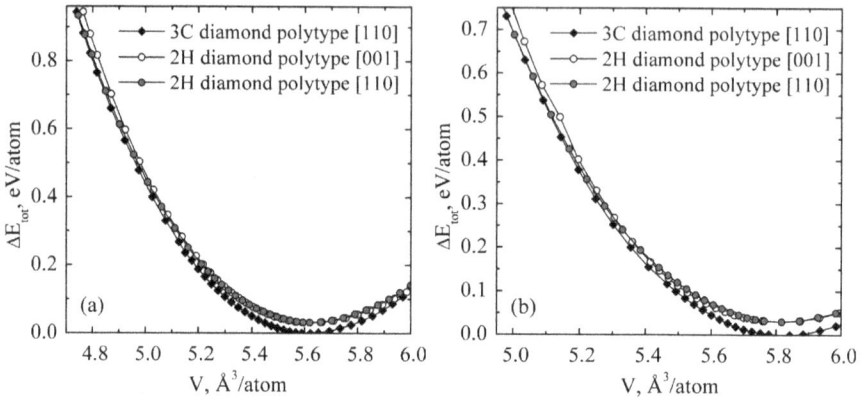

Figure 2. Dependences of difference total energy (ΔE_tot) versus volume (V) for the «3C diamond → 2H diamond» phase transition calculated by the DFT-LDA (a) and DFT-GGA (b) methods.

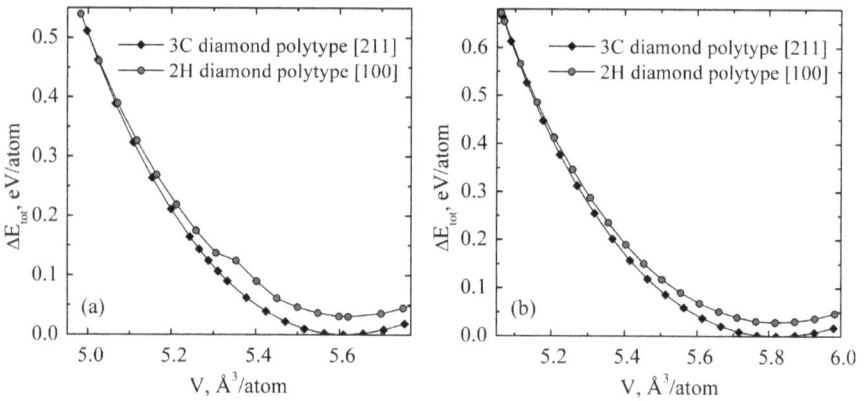

Figure 3. Dependences of difference total energy (ΔE_tot) versus (V) for the «3C diamond → 2H diamond» phase transition calculated by the DFT-LDA (a) and DFT-GGA (b) methods.

Materials Research Forum LLC
doi: http://dx.doi.org/10.21741/9781644900017-29

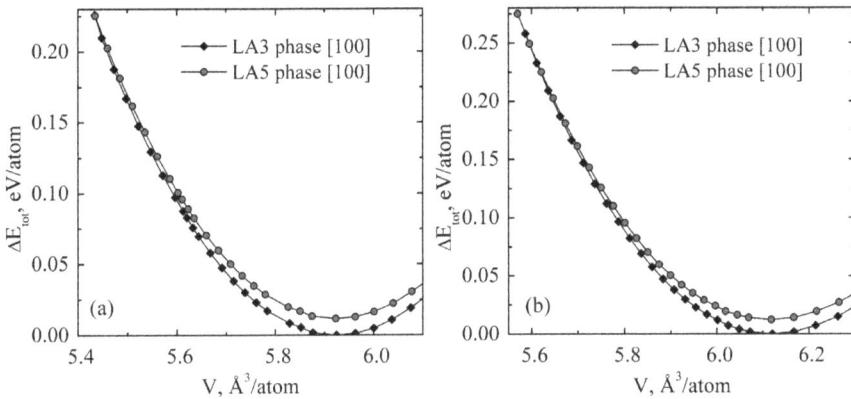

*Figure 4. Dependences of difference total energy (ΔE_{tot}) versus volume (V)
for the «LA3 phase → LA5 phase» phase transition calculated by the DFT-LDA (a)
and DFT-GGA (b) methods.*

Summary

Thus, it was found that martensitic type structural transformations can occur under strong
compression of cubic diamond, as a result of which 2H diamond polytype can be formed (at
pressures from 300 to 380 GPa). Compression of the structure of diamond-like LA3 phase may
cause its transformation into the structure of LA5 phase (at pressures from 160 to 169 GPa).

References

[1] A.G. Khachaturyan, Theory of Structural Transformations in Solids, fhird ed., John Wiley &
Sons, New York, 1983.

[2] D.A. Porter, K.E. Easterling, M. Sherif, Phase Transformations in Metals and Alloys, CRC
Press, Boca Raton, 2009.

[3] E.A. Belenkov, V.A. Greshnyakov, Structure, properties, and possible mechanisms of
formation of diamond-like phases, Phys. Solid State 58 (2016) 2145-2154.
https://doi.org/10.1134/S1063783416100073

[4] J.A. Baimova, L.Kh. Rysaeva, A.I. Rudskoy, Deformation behavior of diamond-like phases:
molecular dynamics simulation, Diam. Relat. Mater. 81 (2018) 154-160.
https://doi.org/10.1016/j.diamond.2017.12.001

[5] V.A. Greshnyakov, E.A. Belenkov, Simulation of the phase transition of graphite to the
diamond-like LA3 phase, Tech. Phys. 61 (2016) 1462-1466.
https://doi.org/10.1134/S1063784216100133

[6] V.A. Greshnyakov, E.A. Belenkov, Investigation on the formation of lonsdaleite from
graphite, J. Exp. Theor. Phys. 124 (2017) 265-274. https://doi.org/10.1134/S1063776117010125

[7] V.A. Greshnyakov, E.A. Belenkov, Modeling of the formation of diamond-like phases from
structural varieties of tetragonal graphite, Letters on Materials 7 (2017) 318-322.
https://doi.org/10.22226/2410-3535-2017-3-318-322

Shape Memory Alloys – SMA 2018   Materials Research Forum LLC
Materials Research Proceedings **9** (2018) 152-156   doi: http://dx.doi.org/10.21741/9781644900017-29

[8] D. Kraus, A. Ravasio, M. Gauthier, D.O. Gericke, J. Vorberger, S. Frydrych, J. Helfrich, L.B. Fletcher, G. Schaumann, B. Nagler, B. Barbrel, B. Bachmann, E.J. Gamboa, S. Gode, E. Granados, G. Gregori, H.J. Lee, P. Neumayer, W. Schumaker, T. Doppner, R.W. Falcone, S.H. Glenzer, M. Roth, Nanosecond formation of diamond and lonsdaleite by shock compression of graphite, Nat. Commun. 7 (2016) 10970. https://doi.org/10.1038/ncomms10970

Shape Memory Alloys – SMA 2018
Materials Research Proceedings 9 (2018) 157-161

Materials Research Forum LLC
doi: http://dx.doi.org/10.21741/9781644900017-30

Ceramic Materials Based on $BaCe_{0.9}M_{0.1}O_{3-\delta}$ for Intermediate Temperature Solid Oxide Fuel Cell

Elena M. Filonenko[1,a], Yuliya A. Lupitskaya[1,b*], Dmitrii A. Kalganov[1,c]

[1]Chelyabinsk State University, 129 Brat'ev Kashirinykh Str., Chelyabinsk 454001, Russia

[a]ponochkachan@gmail.com, [b]lupitskaya@gmail.com, [c]kalganov@csu.ru

*corresponding author

Keywords: Barium Cerate, Heterovalent Substitution, Solid Solutions, Perovskite-Like Phases, Ceramics

Abstract. In the present work, we synthesized compounds based on barium cerate by partial heterovalent substitution of cerium ions Ce^{4+} by ions of yttrium and metals of rare earths (M^{3+} – Pr, Nd, Sm, Gd) on heating in the air. Optimal temperatures for the synthesis of powders were established using data of the thermogravity analysis. The phase features of synthesized compounds have been studied by X-ray diffraction data using the Rietveld method. It is established that at a final synthesis temperature of 1373 K solid solutions of $BaCe_{0.9}M_{0.1}O_{3-\delta}$ are formed, which crystallize within the framework of a distorted perovskite-type structure. For the obtained perovskite phases, the parameters characterizing the structural features were determined, and a correlation between the defectiveness of the crystal lattice of the phase composition and their electrically conductive properties was established. The ceramic properties (relative density, porosity, grain size) of the samples sintered at 1773 K were studied.

Introduction

Oxides of the ABO_3 family with distorted perovskite-like structure is a special class of compounds. Scientific interest in it's due not only to the mechanism of proton transfer [1], but also to the practical application these compounds. In these work solid solutions with a high anionic conductivity [2,3] can be obtained. The mechanism of ion transport in them is achieved by partial substitution of Ce^{4+} ions by low-valent cations in octahedral positions (B-positions) of a perovskite-type structure. As a result of heterovalent substitution in the lattice $AB_{1-x}M_xO_{3-\delta}$, oxygen vacancies with an effective charge of +2 are created and it's concentration $\delta = x/2$ is determined by the level of doping x [2].

Samples of such materials can be obtained by thsintered at the method of polymer complexes (Pechini [3,4]) or by traditional solid state synthesis in the $BaCO_3$-CeO_2-M_2O_3 system (M_{3+} – Y, Pr, Nd, Sm, Gd). To synthesize polycrystalline $BaCe_{1-x}M_xO_{3-\delta}$ samples, chemically pure oxides of CeO_2, Y_2O_3, Pr_6O_{11}, Sm_2O_3, Gd_2O_3, $Nd(OH)_3$ hydroxide and $BaCO_3$ qualification were used by the solid-phase reaction method. The rare earth oxides Pr_6O_{11}, Sm_2O_3, Gd_2O_3 and $Nd(OH)_3$ hydroxide were preliminarily heat treated in air at 1373 K for 1 hour, and barium carbonate was dried at 523 K, respectively.

The prepared mixtures were agitated thoroughly in an agate mortar in accordance with predetermined samples of the initial chemical reagents with the addition of a small amount of ethyl alcohol for 30 minutes. The resulting alcohol-containing batch was pressed under a pressure of 50 MPa in the form of cylindrical samples with a diameter of 14 mm and a thickness of 3–4 mm. Powders of the desired composition were synthesized based on thermogravimetric analysis (TG) data. The composition of the reaction products formed was calculated by weighing on analytical scales to within 0.05 mg. The characteristics of the change in the phase composition of the synthesized samples were studied by the X-ray diffraction analysis (XRD). Identification of the formed compounds was carried out using data from the card catalog of the

Published under license by Materials Research Forum LLC.

Shape Memory Alloys – SMA 2018   Materials Research Forum LLC
Materials Research Proceedings 9 (2018) 157-161   doi: http://dx.doi.org/10.21741/9781644900017-30

International Diffraction Data Center (ICDD JCPDS). Parameters of the crystal lattice of perovskite-type phases were determined using XRD data by the Rietveld method.

The electrical conductivity of the samples was measured in the temperature range from 500 to 975 K with the help of twoprobe method using a R-5083 ac bridge. The ac frequency was 1 kHz and the temperature step was 5 K. The samples had the form of disks 3–4 mm in thickness, with platinum pressure contacts [5].

The morphology of the surface of synthesized ceramic samples of barium cerate and its derivatives was studied using a JEOL JSM-6510 scanning electron microscope.

Results

As follows from the data of thermogravimetric analysis obtained in [3], the process of formation of undoped barium cerate proceeds in one stage, characterized by a high-temperature region (1100–1223 K). X-ray phase analysis data showed that the final product of the synthesis ($T = 1223$ K) is a $BaCeO_3$ compound having a distorted perovskite structure.

In the present work, based on the TG analysis, it was established that the change in the mass of powders of the initial mixtures $BaCO_3$–M_2O_3–CeO_2 (M – Y^{3+}, Pr^{3+}, Nd^{3+}, Sm^{3+}, Gd^{3+}) is completed at 1423 K (Fig. 1).

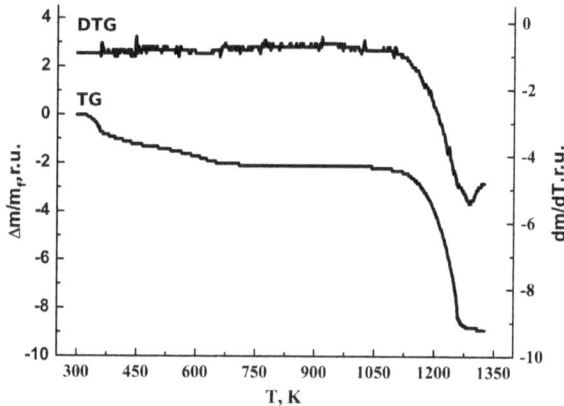

Figure 1. TG- and DTG-thermolysis curves of the initial mixture [$BaCO_3$ + $0.1Y_2O_3$ + $0.9CeO_2$]·nH_2O.

Since the TG analysis was carried out in a dynamic mode, the synthesis temperatures of the materials were slightly reduced and amounted to 1373 K with a holding time of 3 hours.

The XRD results for the synthesized $BaCe_{0.9}M_{0.1}O_{3-\delta}$ samples are shown in Fig. 2. As can be seen from the figure, the X-ray diffraction patterns of the compounds under study contain the same set of diffraction peaks, the shape and half-width and the number of which do not vary in a given range of Bragg angles, indicating that the symmetry type of the lattice of the phases formed isostructural with orthorhombic $BaCeO_3$.

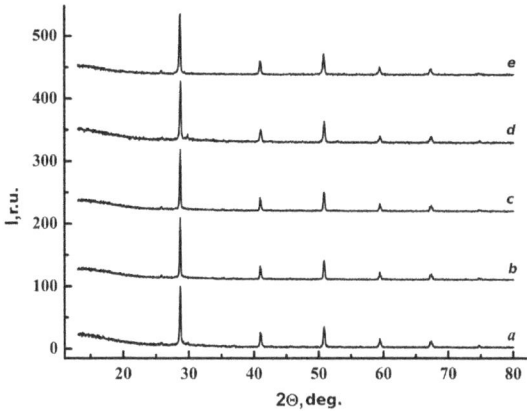

Figure 2. Diffractograms of samples $BaCe_{0.9}M_{0.1}O_{3-\delta}$, sintered at 1373 K:
a – $BaCe_{0.9}Y_{0.1}O_{3-\delta}$, b – $BaCe_{0.9}Pr_{0.1}O_{3-\delta}$, c – $BaCe_{0.9}Nd_{0.1}O_{3-\delta}$,
d – $BaCe_{0.9}Sm_{0.1}O_{3-\delta}$, e – $BaCe_{0.9}Gd_{0.1}O_{3-\delta}$.

The refinement of the crystal lattice parameters for barium cerate compounds, as noted above, was carried out by the Rietveld method using the GSAS software. The convergence of theoretical curves with experimental data was evaluated by Bragg and structural factors, which did not exceed 1.64 and 2.10 for a perovskite-type structure with orthorhombic distortion (Fig. 3).

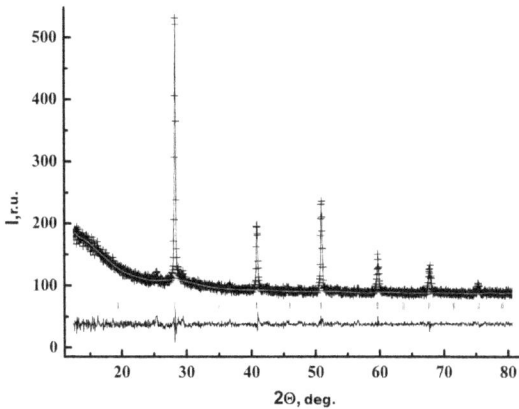

Figure 3. Experimental, theoretical and difference diffractograms
of the phase of composition $BaCe_{0.9}Nd_{0.1}O_{3-\delta}$.

Analysis of X-ray diffraction data allowed us to conclude that for pseudocubic phases the values of the parameter a and the volume V_a of the unit cell directly depend on the ionic radius of the dopant (Table 1).

Table 1. Values of the ionic radii of the dopant, parameter, volume of the unit cell, conductivity, and activation energy of the conductivity for phases of composition $BaCe_{0.9}M_{0.1}O_{3-\delta}$.

Composition	M^{3+}	r, [Å]	a, [Å]	V_a, [Å^3]	σ, [S/m]	E_a, [kJ/mol] (500–973 K)
$BaCeO_3$	-	-	4.377	83.855	$2.6\ 10^{-4}$	88±4
$BaCe_{0.9}Y_{0.1}O_{3-\delta}$	Y	0.87	4.372	83.568	$1.2\ 10^{-4}$	85±4
$BaCe_{0.9}Pr_{0.1}O_{3-\delta}$	Pr	1.02	4.409	85.708	$8.1\ 10^{-3}$	77±4
$BaCe_{0.9}Nd_{0.1}O_{3-\delta}$	Nd	0.99	4.403	85.358	$7.6\ 10^{-3}$	75±4
$BaCe_{0.9}Sm_{0.1}O_{3-\delta}$	Sm	0.97	4.397	85.010	$6.3\ 10^{-3}$	73±4
$BaCe_{0.9}Gd_{0.1}O_{3-\delta}$	Gd	0.94	4.396	84.952	$5.6\ 10^{-3}$	74±4

Partial heterovalent substitution of ions of yttrium and metals of rare earths ($M - Pr^{3+}$, Nd^{3+}, Sm^{3+}, Gd^{3+}) in $BaCe_{0.9}M_{0.1}O_{3-\delta}$ phases leads to a monotonous increase in the parameter a and volume Va of the unit cell, respectively (Table 1). The observed increase in the structural parameters leads to a significant distortion of the crystal lattice, which contributes to the formation of oxygen vacancies, and, consequently, to the increase in conductivity in synthesized compounds (Table 1).

Ceramic samples made in the form of tablets were prepared by pressing the powders followed by sintering at 1773 K for 3 hours (Fig. 4).

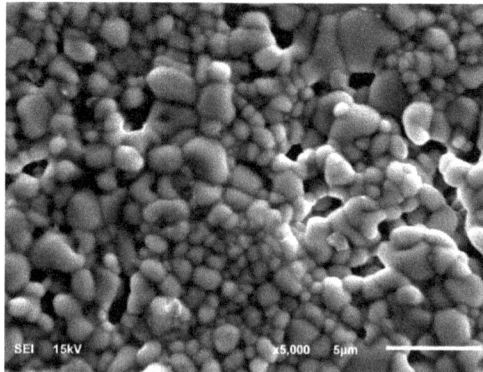

Figure 4. Microscopic image of the surface of the sample $BaCe_{0.9}Gd_{0.1}O_{3-\delta}$, sintered at 1773 K.

Depending on the composition, all the materials under study have a high relative density (91.2–95.6 %). To qualitatively evaluate the microstructure of the surface of the sintered ceramics, the dimensions of each particle were determined, on the basis of which the histograms of the particle distribution were plotted and their mean size, which was 2.7 μm, was calculated.

It should be noted that such materials are characterized by high thermodynamic and kinetic stability, which makes them promising solid electrolytes for IT-SOFC.

Summary

Thus, for the compounds of barium cerate, formed by partial heterovalent substitution of Ce^{4+} ions of low-valence metals ($M - Y^{3+}$, Pr^{3+}, Nd^{3+}, Sm^{3+}, Gd^{3+}) in the $BaCO_3–M_2O_3–CeO_2$ system under heating in air, the conditions of preparation were studied. The analysis of X-ray diffractograms showed that $BaCe_{0.9}M_{0.1}O_{3-\delta}$ solid solutions crystallize within the framework of a perovskite-type structure characterized by orthorhombic distortion (the space group Pmcn). With the help of scanning electron microscopy, the morphology of the surface of ceramic materials of various compositions was studied, the relative density was determined, and the average grain size was calculated.

Acknowledgments

The work was supported by the Russian Foundation for Basic Research in the scientific project No. 18-33-00269

References

[1] M. Amsif, D. Marrero-Lopez, J.C. Ruiz-Morales, Influence of rare-earth doping on the microstructure and conductivity of $BaCe_{0.9}Ln_{0.1}O_{3-\delta}$ proton conductors, J. Power Sources 196 (2011) 3461-3469. https://doi.org/10.1016/j.jpowsour.2010.11.120

[2] D. Medvedev, A. Murashkina, $BaCeO_3$: Materials development, properties and application, Prog. Mater. Sci. 60 (2014) 72-129. https://doi.org/10.1016/j.pmatsci.2013.08.001

[3] Yu.A. Lupitskaya, E.M. Filonenko, D.A. Kalganov, Synthesis and ionic conductivity of compounds with partial substitution of cerium by rare-earth ions in barium cerate, Bulletin of Chelyabinsk State University. Series "Phisiks" 21 (2015) 143-147.

[4] L.W. Tai, P.A. Lessing, Modified resin-intermediate processing of perovskite powders. Part I. Optimization of polymeric precursors, J. Mater. Res. 7 (1992) 502-510. https://doi.org/10.1557/JMR.1992.0502

[5] Yu.A. Lupitskaya, D.A. Kalganov, M.V. Klyueva, Formation of Compounds in the Ag_2O-Sb_2O_3-MoO_3 System on Heating, Inorg. Mater. 54 (2018) 240-244. https://doi.org/10.1134/S0020168518030081

Shape Memory Alloys – SMA 2018
Materials Research Proceedings **9** (2018) 162-166

Materials Research Forum LLC
doi: http://dx.doi.org/10.21741/9781644900017-31

The Structural Phase Diagrams of Fe-*Y* (*Y* = Ga, Ge, Al) Alloys

Mariya V. Matyunina[1,a*], Mikhail A. Zagrebin[1,2,b], Vladimir V. Sokolovskiy[1,2,c], Vasiliy D. Buchelnikov[1,2,d]

[1]Chelyabinsk State University, 129 Brat'ev Kashirinykh Str., Chelyabinsk 454001, Russia

[2]National University of Science and Technology "MISiS", 4 Leninskiy Prospect, Moscow 119991, Russia

[a]matunins.fam@mail.ru, [b]miczag@mail.ru, [c]vsokolovsky84@mail.ru, [d]buche@csu.ru

*corresponding author

Keywords: *Ab Initio* Calculations, Phase Transformations, Phase Diagram, Fe-Ga, Fe-Al, Fe-Ge

Abstract. This work presents a theoretical study of structural and magnetic properties from first-principles calculations for binary $Fe_{100-x}Y_x$ type alloys (Y = Al, Ga, Ge) in concentration range $18.75 \leq x \leq 31.25$ at.%. It is shown that as chemical disorder increases, the lattice parameters are found to increase as opposed to the magnetic moment behavior as a function of Al, Ga, Ge content for all investigated phases of binary Fe-*Y* type alloys. The phase diagrams obtained by comparing the computed total energies between all phases studied are in agreement with the experimental results.

Introduction

Nowadays, promising multifunctional materials based on α-Fe such as Fe-Al, Fe-Ga, and Fe-Ge alloys are in focus of interest owing to their unusual mechanical, magnetic and electrical properties [1-4]. The Fe-Ga alloys are successful magnetostrictive materials, which demonstrated two peaks of saturation magnetostriction λ_{001} 395×10^{-6} and 350×10^{-6} at room temperature for compositions with $x \approx 19$ at.% and $x \approx 27$ at.%, respectively [1]. The Fe-Al alloys are attractive due to their low density, high elastic modulus, low cost and outstanding high-temperature corrosion resistance in oxidizing and environments [2,3]. The system of Fe-Ge has been interesting due to their potential applications in permanent magnets and magnetic refrigeration, in particular, it is a good ferromagnet at room temperature with the Curie temperature about 640 K [4]. These systems are less studied as compared with previously considered Fe-Ga and Fe-Al alloys [3].

The aim of this work is a study of the magnetic and structural properties of $Fe_{100-x}Y_x$ (Y = Al, Ga, Ge) alloys using the density functional theory at zero temperature.

Computation details

The first-principles calculations of structural and magnetic properties of the Fe-*Y* alloys were done using the density functional theory which implements in Vienna *Ab-initio* Simulation Package (VASP) [5,6]. The exchange-correlation coupling was treated with the generalized gradient approximation (GGA) in Perdew-Burke-Ernzerhof (PBE) form [7]. For the pseudopotentials used, the electronic configurations were Fe ($3p^6 3d^7 4s^1$), Al ($3s^2 3p^1$), Ga ($3d^{10} 4s^2 4p^1$) and Ge ($3d^{10} 4s^2 4p^2$). The k-point mesh was generated by the Monkhorst-Pack scheme [8] with a grid of 8×8×8 points. The value of the kinetic energy cut-off for the augmentation charges is 800 eV, the plane-wave cut-off energy is 400 eV. All calculations were done at zero temperature (T = 0 K).

Published under license by Materials Research Forum LLC.

Shape Memory Alloys – SMA 2018   Materials Research Forum LLC
Materials Research Proceedings **9** (2018) 162-166   doi: http://dx.doi.org/10.21741/9781644900017-31

The geometric optimization was done with an account of the 32 atoms supercell approach for the following phases:

the A2 with an α-Fe-type structure with Fe and Ga atoms randomly distributed, space group Im-3m;

the B2 with a CsCl-type structure with Fe and Ga atoms partially ordered, space group Pm-3m;

the $D0_3$ with a BiF_3-type structure with Fe and Ga atoms partially ordered, space group Fm-3m;

the $D0_{19}$ with a $MgCd_3$-type structure with Fe and Ga atoms partially ordered, space group P63/mmc;

the $L1_2$ with a Cu_3Au-type structure with Fe and Ga atoms partially ordered, space group Pm-3m.

According to the experimental phase diagrams [9] for Fe-Al alloys were considered A2, B2, and $D0_3$ structures, and for Fe-Ga (Ge) alloys calculations were done for five types of ordering. To create the off-stoichiometric composition, the one type of atoms (Fe) has been replaced by another type (Al, Ga, Ge) one. Thus, the 32-atom supercell allows us to change the composition by the step of 3.125 at.%.

Results and Discussion

Let us discuss the obtained results. We started from calculations of geometric structure optimization of supercells. The Fig. 1 shows the calculation results of atomic volume of considered phases as function of Al (Ga, Ge) content (at.%) for $Fe_{100-x}Y_x$ (Y = Al, Ga, Ge) alloys. To obtain the atomic volume following equation was used: $V_a = V_c / N$, where V_c is the volume of the elementary cell and N is the number of atoms per cell ($N_{A2, B2} = 2$, $N_{D03} = 16$, $N_{D019} = 8$). It is seen that a difference between atomic volumes of each phase appears with increasing adding elements concentration. For the range of investigation ($18.75 \leq x \leq 31.25$ at.%), the volume change between disordered A2 and ordered $D0_3$ phases is visible for all type alloys. Notice, that for Fe-Ge alloys this difference is about twice in comparing with other Fe-Ga and Fe-Al alloys. The values of atomic volume for $D0_{19}$ and $L1_2$ phases (Figs. 1(b), c)) are increasing with increasing Ga (Ge) content for Fe-Ga (Fe-Ge) alloys, however, the difference between these values stay almost constant and draw about 0.08 Å^3 (0.05 Å^3) respectively.

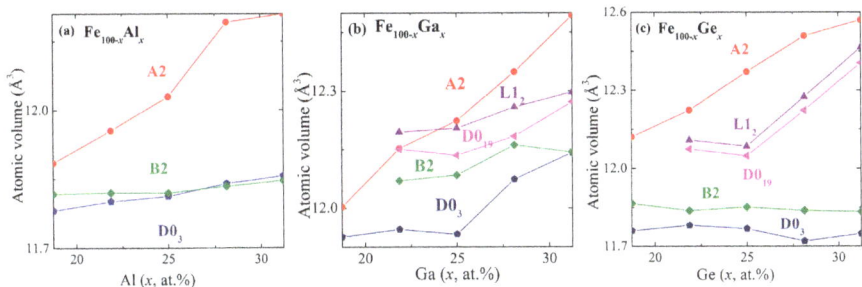

Figure 1. The calculated atomic volume of considered phases in dependence on (a) Al, (b) Ga, (c) Ge concentration (at.%) for $Fe_{100-x}Y_x$ (Y = Al, Ga, Ge) alloys.

The composition dependences of total magnetic moment per atom for all studied compositions and considered phases are presented in Fig. 2. It is observed that the addition of Al (Fig. 1(a)), Ga (Fig. 1(b)) and Ge (Fig. 1(c)) to iron results in a decrease in the total magnetization due to the dilution of the magnetic subsystem by non-magnetic doping atoms.

Shape Memory Alloys – SMA 2018   Materials Research Forum LLC
Materials Research Proceedings 9 (2018) 162-166   doi: http://dx.doi.org/10.21741/9781644900017-31

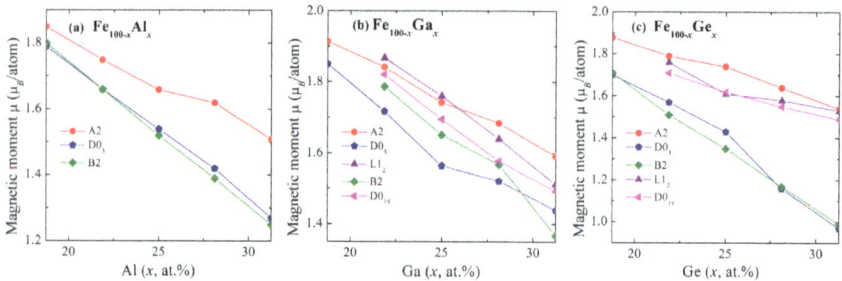

Figure 2. The calculated total magnetic moment of considered phases in dependence on (a) Al,
(b) Ga, (c) Ge concentration (at.%) for $Fe_{100-x}Y_x$ (Y = Al, Ga, Ge) alloys.

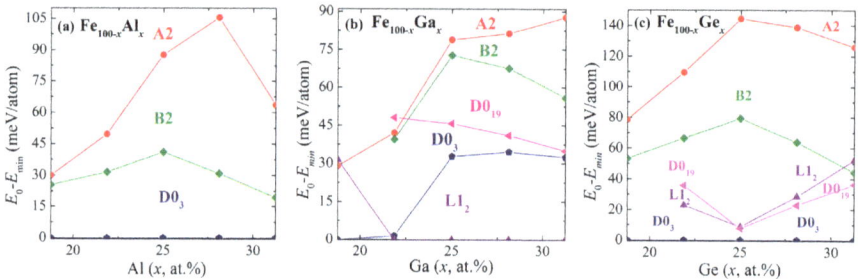

Figure 3. The calculated energy difference of A2, $D0_3$, B2, $L1_2$, and $D0_{19}$ structures with respect
to the minimum energy of structures as a function of (a) Al, (b) Ga, (c) Ge concentration
for $Fe_{100-x}Y_x$ (Y = Al, Ga, Ge) alloys. Here, E_0 and E_{min} are the ground state energy and the
energy of a favorable phase, respectively.

The Fig.3 shows the calculated phase diagrams for $Fe_{100-x}Y_x$ (Y = Al, Ga, Ge) alloys in the
range $18.75 \leq x \leq 31.25$ at.%. To plot these phase diagrams, we compared the energy difference,
$E_0 - E_{min}$ (where E_0 is a ground state energy, E_{min} is the energy of a favorable phase) between A2,
B2, and $D0_3$ structures for Fe-Al alloys (Fig. 3(a)), and between A2, B2, $D0_3$, $L1_2$, and $D0_{19}$ for
Fe-Ga (Fig. 3(b)) and Fe-Ge (Fig. 3(c)) alloys. In Fig. 3(a), E_{min} corresponds ground state energy
of $D0_3$ in the range $18.75 \leq x \leq 31.25$ at.%. The sequence of phase transitions between ordered
($D0_3$) and fully disordered (A2) structures realize in two steps: $D0_3 \rightarrow B2 \rightarrow A2$. In range
$x = 22$–35 at.% at temperatures low than 600 °C this sequence of phase transitions observes
experimentally [9,10].

For the phase diagram of Fe-Ga alloys presented in Fig. 3(b), E_{min} corresponds to a $D0_3$
structure in the range $18.75 \leq x < 21.875$ at.%, and describes follows phase transition:
$D0_3 \rightarrow L1_2 \rightarrow A2$. In the range $21.875 \leq x \leq 31.25$ at.% E_{min} corresponds to an $L1_2$ structure, and
the sequence of transitions between ordered ($L1_2$) and fully disordered (A2) phases realize in
three steps: $L1_2 \rightarrow D0_3 \rightarrow D0_{19} \rightarrow B2 \rightarrow A2$. The recent in situ neutron diffraction study of Fe-
Ga alloys performed by Golovin et al. [11] has shown that compounds with $x = 27$ at.% possess
the $D0_3$ structure after casting and the first-order $L1_2$ to $D0_{19}$ transition takes place at 620 °C.

For the third phase diagram of Fe-Ge alloys, which depicted in Fig. 3(c), it can be observed
that the $D0_3$ phase is the favorable phase in the considered range of investigation. In the region of
concentration $18.75 \leq x \leq 25$ at.% is realized the following sequence of phase transitions

$D0_3 \rightarrow L1_2 \rightarrow D0_{19} \rightarrow B2 \rightarrow A2$ and for the region $25 < x \leq 31.25$ at.% the sequence changes to the $D0_3 \rightarrow D0_{19} \rightarrow L1_2 \rightarrow B2 \rightarrow A2$. According to the experimental study of $Fe_{77.5}Ge_{22.5}$ by A. Fernandez et al. [12] at 655 °C or lower temperatures is observed the Ll_2 super-lattice while at 665 °C or higher temperatures the $D0_{19}$ superlattices is realized. The phase change $D0_{19} \rightarrow L1_2$ takes place at around 660 °C. This temperature dependence was observed by Belamri et al. [13] for air-cooled Fe-23Ge alloy, and shown an anomaly between 150 °C and 950 °C: a contraction at 350 °C due to the $D0_3 + B8_1 \rightarrow D0_3 + L1_2$ transition, followed by an expansion linked to the $D0_3 + L1_2 \rightarrow D0_3 + D0_{19}$ transition.

Summary

In this work, we have introduced first-principles calculations to study the structural and magnetic properties of $Fe_{100-x}Y_x$ (Y = Al, Ga, Ge) alloys in the range $18.75 \leq x \leq 31.25$ at.%. The geometry optimization of A2, $D0_3$, and B2 structures for Fe-Al alloys, and A2, $D0_3$, B2, $L1_2$, and $D0_{19}$ structures for Fe-Ga, Fe-Ge has been performed by using the VASP package. It is shown that for all types of alloys and phases the atomic volumes rises while the total magnetic moment reduces with increasing adding elements content (Al, Ga, Ge). Based on the total energy calculations at zero temperature ($T = 0$ K) we obtained the phase diagrams of considered alloys and compositions in term of the energy scale. The scale of energy difference ($E_0 - E_{min}$) which we used to plot phase diagrams, is proportional to the transition temperature (T_0) via a rough approximation $\Delta E \approx k_B T_0$. In our case, the temperature of phase transitions is lower than the experimental data, nevertheless, a good qualitative agreement between the theoretical sequence of phase transitions and experimental ones [9,13] can be seen.

Acknowledgments

This work was supported by Russian Science Foundation No. 18-12-00283.

References

[1] A.E. Clark, K.B. Hathaway, M. Wun-Fogle et al., Extraordinary magnetoelasticity and lattice softening in bcc Fe-Ga alloys, J. Appl. Phys. 93 (2003) 8621-8623. https://doi.org/10.1063/1.1540130

[2] L.A. Talischi and A. Samadi, Effect of Stress-Induced Phase Transformation on the Fracture Toughness of Fe_3Al Intermetallic Reinforced with Yttria-Partially Stabilized Zirconia Particles, Metall. Mater. Trans. A, DOI: 10.1007/s11661-017-4234-3. https://doi.org/10.1007/s11661-017-4234-3

[3] I.S. Golovin, A.M. Balagurov, I.A. Bobricov, J. Cifre, Structure induced anelasticity in Fe_3Me (Me = Al, Ga, Ge) alloys, J. Alloys Compd. 688 (2016) 310-319. https://doi.org/10.1016/j.jallcom.2016.06.277

[4] K.V. Shanavas, M.A. McGuire, D.S. Parker, Electronic and magnetic properties of Si substituted Fe_3Ge, J. Appl. Phys. 118 (2015) 123902. https://doi.org/10.1063/1.4931574

[5] G. Kresse, J. Furthmüller, Efficient iterative schemes for ab initio total-energy calculations using a plane-wave basis set, Phys. Rev. B 54 (1996) 11169-11186. https://doi.org/10.1103/PhysRevB.54.11169

[6] G. Kresse, D. Joubert, From ultrasoft pseudopotentials to the projector augmented-wave method, Phys. Rev. B 59 (1999) 1758-1775. https://doi.org/10.1103/PhysRevB.59.1758

[7] J.P. Perdew, K. Burke, M. Enzerhof, Generalized Gradient Approximation Made Simple, Phys. Rev. Lett. 77 (1996) 3865-3868. https://doi.org/10.1103/PhysRevLett.77.3865

[8] H.J. Monkhorst, J.D. Pack, Special points for Brillouin-zone integrations, Phys. Rev. B. 13 (1976) 5188-5192. https://doi.org/10.1103/PhysRevB.13.5188

[9] O. Kubaschewski, Iron-binary Phase Diagrams, Springer-Verlag, Berlin, 1982.

[10] M. Palm, Concepts derived from phase diagram studies for the strengthening of Fe–Al-based alloys, Intermetallics 13 (2005) 1286-1295. https://doi.org/10.1016/j.intermet.2004.10.015

[11] I.S. Golovin, A.M. Balagurov, V.V. Palacheva, A. Emdadi, I.A. Bobrikov, V.V. Cheverikin, A.S. Prosviryakov, S. Jalilzadeh, From metastable to stable structure: the way to construct functionality in Fe-27Ga alloy, J. Alloys Compd. 751 (2018) 364-369. https://doi.org/10.1016/j.jallcom.2018.04.127

[12] A. Fernandez, L. Tejedor, L. Bru, Electron microscopy study of the phase change $D0_{19} \rightarrow L1_2$ in the Fe_3Ge compound, Phys. Stat. Sol. (a) 34 (1976) K17-K19. https://doi.org/10.1002/pssa.2210340147

[13] Z. Belamri, D. Hamana, I.S. Golovin, Study of order-disorder transitions in Fe-Ge alloys and related anelastic phenomena, J. Alloys Compd. 554 (2013) 348-356. https://doi.org/10.1016/j.jallcom.2012.11.012

Shape Memory Alloys – SMA 2018
Materials Research Proceedings **9** (2018) 167-173

Materials Research Forum LLC
doi: http://dx.doi.org/10.21741/9781644900017-32

On Thermodynamic Description of Finite-Size Multiferroics

Ivan A. Starkov[1,2,a*], Abdulkarim A. Amirov[3,4,b], Alexander S. Starkov[2,c]

[1]Nanotechnology Center, St. Petersburg Academic University, 8/3 Khlopin Str., St. Petersburg 194021, Russia

[2]National Research University of Information Technologies, Mechanics and Optics (ITMO University), 49 Kronverksky Prospect, St. Petersburg 197101, Russia

[3]Center for Functionalized Magnetic Materials (FunMagMa) & Institute of Physics Mathematics and Informational Technologies Immanuel Kant Baltic Federal University, 14 A. Nevskogo Str., Kaliningrad 236041, Russia

[4]Amirkhanov Institute of Physics, Daghestan Scientific Center, Russian Academy of Sciences, 94 Yagarskogo Str., Makhachkala 367003, Russia

[a]starkov@spbau.ru, [b]amiroff_a@mail.ru, [c]ferroelectrics@ya.ru

*corresponding author

Keywords: Multiferroics, Boundary Conditions, Variational Principles, Multicaloric Effect

Abstract. We present an accurate description of the thermodynamic processes occurring in finite-size multiferroics. Our approach avoids the errors that stem from a misunderstanding of the basics of differential and variational calculus. That is, a rigorous formulation of the problem taking into account the presence of gradient terms is presented. The developed theoretical framework provides analysis of a multiferroic layer with consideration of flexoelectric effect.

Introduction

A change in temperature or/and entropy of the solid caused by the application or removal of the external field corresponds to a caloric effect (CE). Apparently, forces of different nature can induce a variety of such phenomena. They are named according to the field that causes them, namely, the electro-, magneto-, and elastocaloric. If the solid possesses more than one CE, then such a material is said to be a multicaloric. It is necessary to mention, that a lot of materials are multicalorics, however, a noticeable multicaloric effect, i.e. sufficient for practical applications, is nowadays only known in multiferroics. The prospect of using them to create high environmentally friendly cooling systems explains the growing interest of researchers in recent years. Unfortunately, the rising popularity and complexity of the physics behind thermodynamic processes have led to a series of erroneous papers in leading journals. The number of mistakes in the differential equations that describe the state of a multiferroic is relatively small. At the same time, the boundary conditions (BCs) for multiferroic materials had been considered only partially, and that results in a significant amount of errors in their formulation. This work aims to provide a comprehensive thermodynamic approach for modeling of finite-size multiferroics.

Main equations

Usually, the free energy density w is used for the thermodynamic description of the multiferroic material. This quantity is characterized by generalized forces x_i, $i = 1, 2, ..., n$ and generalized coordinates X_i. The first ones are independent and the latter are dependent variables (functions). In some sources, X_i are also called generalized displacements. The variation of the total energy functional W is performed with respect to these variables that must be independent among themselves. Following the standard procedure, we introduce the generalized vector of displacement **X** with the components X_i. In addition, we define the tensors $\nabla \mathbf{X}$ and $\nabla\nabla \mathbf{X}$ with the components $X_{i,j}$ and $X_{i,jk}$, respectively (i.e. the first and the second gradients of the

Published under license by Materials Research Forum LLC.

Shape Memory Alloys – SMA 2018   Materials Research Forum LLC
Materials Research Proceedings **9** (2018) 167-173   doi: http://dx.doi.org/10.21741/9781644900017-32

displacement). Any index after a comma will hereafter indicate a partial differentiation with respect to the corresponding coordinate $X_{i,j} = \partial X_i/\partial x_j$. We assume that the energy density w depends only on the generalized displacements and their derivatives up to second order. Nevertheless, it easy to verify that the form of the dependence w on x_i does not affect the subsequent analysis. Thus, the total energy W stored in the volume V is

$$W \equiv \int_V w(\mathbf{X}, \nabla\mathbf{X}, \nabla\nabla\mathbf{X})\mathrm{d}V. \tag{1}$$

Similar functionals arise in the ordinary theory of elasticity [1], as well as in the description of multiferroics in linear [2] and non-linear approximation [3]. The functionals containing derivatives of higher order than in Eq.1 appear in the consideration of rods and shells. The boundary conditions for them can be obtained by the scheme described below. Nevertheless, their analysis is not currently possible as the number of BCs for the multiferroic shells may be close to a hundred. In view of this, we restrict our study by the functional Eq. 1, which involves derivatives up to the second order. Let us investigate this functional.

Choice of generalized coordinates

The choice of the independent variables plays an important role and can cause a possible systematic error. However, these errors are rather methodical and allow to write the correct equations, but not to derive it. To be more precisely, the wrong choice of variables would lead to incorrect results while considering the boundary problem. When describing electro-magneto-elastic phenomena, the vectors of the electric and magnetic fields E_i, H_i, $i = 1, 2, 3$, together with the strain u_{ij} or stress σ_{ij} tensors, are usually chosen as independent variables [4]. In turn, the thermodynamically conjugate quantities, i.e. the polarization P_i, the magnetization M_i, and the stress or strain tensors are treated as dependent variables. Recall that the strain tensor associated with the displacement vector u_i is given by the relations $u_{ij} = (u_{i,j} + u_{j,i})/2$. This way, the only independent variables are three displacements u_i. The six elements of the stress σ_{ij} or strain u_{ij} tensors turn out to be dependent variables. For instance, the deformation components are connected by the additional relations [1]

$$u_{ik,lm} + u_{lm,ik} = u_{il,km} + u_{km,il}. \tag{2}$$

Identical relations hold for the stress tensor. Therefore, we can choose them as independent variables only when we know that the equalities Eq. 2 are fulfilled (e.g., if u_{ik} or σ_{ij} are constants). Otherwise, it is necessary to find the conditional extremum Eq. 1 with three auxiliary conditions. Thus, the use of the strain u_{ij} or stress σ_{ij} tensors as independent variables is unacceptable in the deliberately non-uniform electric or elastic field. As an illustration of the possible errors arising from the wrong choice of variables, we consider a special case of the functional in Eq. 1. Assume that it depends only on the strain tensor $w = w(u_{ij})$. The variation of Eq. 1 with respect to independent set u_i leads to the classical equations of elasticity

$$\sigma_{ij,j} = 0, \tag{3}$$

where $\sigma_{ij} = \partial w/\partial u_{ij}$. Here the Einstein summation convention is used, in which repeated indexes are summed over. Meanwhile, a standard procedure for the derivation of the Eq. 3 used in [4] is as follows. The functional $\hat{w} = w(u_{ij}) - u_{ijij}$ is considered instead of w. The density $\hat{w}$ is varied while u_{ij} are taken as independent variables. As a result, we obtain the definition of stress tensor, and the Eq. 3, in fact, are postulated. Because of the correct consideration of the equations, the choice of u_{ij} as independent variables is a methodological mistake. Put simply, in this case the

Shape Memory Alloys – SMA 2018   Materials Research Forum LLC
Materials Research Proceedings **9** (2018) 167-173   doi: http://dx.doi.org/10.21741/9781644900017-32

equations arise from out of nowhere. If we consider the entropy density $S = -\partial w/\partial T$, where T is temperature, then we already get different expressions for the energy density w and $\hat{w}$.

A similar problem arises in the description of the electric and magnetic fields. The functionals Eq.1 with densities $w_\varphi = \varphi_{,i}\varphi_{,i}$ and $w_E = E_i E_i$ take the same values for the established relationship between the electric potential and electric field strength $Ei = -\varphi_{,i}$. But the functional written in terms of E_i cannot provide any equation. At the same time, the functional w_φ corresponds to the classical Laplace equation. Using w_E instead of the correct w_φ is only legitimate when the electric field strength E_i is known in advance. In multiferroics, the electric, magnetic, and elastic fields are coupled with each other, i.e. it is not possible to determine one of them without finding another. Therefore, the inclusion of w_E to the full functional Eq. 1 can be considered just as an approximation. The aforementioned facts allow us to draw an important conclusion. In the derivation of the equations of state, the variation should be performed with respect to the independent variables, namely, the electric (φ) and magnetic (ψ) potentials and displacements u_i. The variation of Eq. 1 with respect to the components of the electric (E_i) and magnetic fields H_i, as well as the strain u_{ij} or stress σ_{ij} tensors can induce serious errors.

General theory of boundary conditions
The need to introduce additional boundary conditions is discussed more in [5]. Nonetheless, in the cited paper, the equations and BCs are derived from the different physical principles and are not consistent with each other. The conditions are given either partially or not at all. In contrast to the usual theory of elasticity or piezoelectricity, we need to consider higher order derivatives in Eq. 1 for multiferroics. This leads to an increase in the number of boundary conditions. Another way is to replace BCs with the standard conditions for elasticity and electro/magnetostatics [6]. In view of this, it is necessary to understand what the boundary conditions should be used to describe the multiferroic by taking into account the gradient terms in Eq.1. To do this, first we give a detailed derivation of the boundary conditions in the general case, then describe the equations of electro-magneto-elasticity. And only afterwards proceed to the discussion of inaccuracies and errors in the BCs.

The present method for obtaining the boundary conditions is quite general and based on the analysis of the total variation of the free energy density. It is based on a generalization used in [7] for the equations of elasticity and easier than the fiducial approach of [8]. The variation of Eq.1 with respect to X_i gives

$$\delta W = \int_V (\rho_i \delta X_i + \sigma_{ij} \delta X_{i,j} + \tau_{ijk} \delta X_{i,jk})\,\mathrm{d}V, \tag{4}$$

where $\{\rho_i,\ \sigma_{ij},\ \tau_{ijk}\}$ are the variables thermodynamically conjugated to X_i, $X_{i,j}$, and $X_{i,jk}$, respectively.

$$\rho_i = \frac{\partial w}{\partial X_i}, \quad \sigma_{ij} = \frac{\partial w}{\partial X_{i,j}}, \quad \tau_{ijk} = \frac{\partial w}{\partial X_{i,jk}}. \tag{5}$$

These variables are also called conventional stress (σ) and higher stress (τ) [7]. Using the divergence (Gauss-Ostrogradsky) theorem, we can convert the volume integrals in Eq. 4 as

$$\delta W = \int_V \left[(\rho_i - \sigma_{ij,j} + \tau_{ijk,jk})\delta X_i\right]\mathrm{d}V + \int_S [(\sigma_{ij} - \tau_{ijk,i})n_j\delta X_i + \tau_{ijk}n_k\delta X_{i,j}]\mathrm{d}S, \tag{6}$$

where S is the surface bounding the volume V, n_j are the components of the normal vector to S. The extremality of Eq. 1 according to (6) provides the equilibrium relation

$$\rho_i - \varsigma_{jk,j} = 0. \tag{7}$$

Here, the generalized stress ς_{jk} is defined as $\varsigma_{jk} = \sigma_{jk} - \tau_{ijk,i}$. The expression Eq. 7 is the Euler-Lagrange equation [9].

Up to this point, we have bypassed the basic ideas of the calculus of variations. To obtain the BCs for Eq. 1, it is necessary to go beyond the existing theoretical framework and generalize the results of [7] to the case of nonlinear dependence w for an arbitrary number of variables $\{X_i, X_{i,j}, X_{i,jk}\}$. It is to be mentioned that the author of [7] considers linear equations of the theory of elasticity, and we plan to explore the electro-magneto-elastic equation with gradient terms included. The quantities $\delta X_{k,j}$ cannot be considered as independent of δX_k on the surface S because, if δX_k is known on S, so is the surface gradient of δX_k. Therefore, the correct formulation of the problem demands $2n$ BCs, e.g. prescribed values for δX_k and their normal derivatives. In order to identify independent boundary conditions, we represent the derivative $\delta X_{k,j}$ as

$$\delta X_{k,j} = d^{\|}_j \delta X_k + n_j d^{\perp} \delta X_k, \tag{8}$$

i.e. decompose the derivative into its tangential $d^{\|}_j$ and normal $n_j d^{\perp}$ components

$$d^{\perp} \equiv n_k \frac{\partial}{\partial x_k}, \quad d^{\|}_j \equiv (\delta_{jk} - n_j n_k) \frac{\partial}{\partial x_k}, \tag{9}$$

with δ_{jk} as the Kronecker delta. The substitution of Eq. 8 into Eq. 6 and accounting Eq. 7 with the surface divergence theorem [7] yield

$$\delta W = \int_S (T_k \delta X_k + R_k d^{\perp} \delta X_k) \, dS. \tag{10}$$

Here we have introduced the notations

$$T_i \equiv n_k \varsigma_{ik} + n_j n_k \tau_{ijk}(d^{\|}_l n_l) - d^{\|}_j(n_k \tau_{ijk}), \quad R_i \equiv n_j n_k \tau_{ijk}, \tag{11}$$

for the generalized surface traction T_k and double stress traction R_k. The variation of W must vanish, i.e., as can be seen from the Eq. 10, the following boundary conditions must be fulfilled at the interfaces

$$[X_k] = 0, \quad [d^{\perp} X_k] = 0, \quad [T_k] = 0, \quad [R_k] = 0, \tag{12}$$

where the symbol $[X]$ denotes the jump of X across the interface. The last two terms in Eq. 12 describe so-called natural BCs [9].

To sum up, the field with the energy density w has to satisfy n equations Eq. 7. The amount of BCs depends on the number of second derivatives of the generalized displacements X. When there are second derivatives $X_{i,jk}$ for every i, this number is equal to $4n$ for the conditions at the inner boundary of the media Eq. 12. For traction-free interfaces, there have to be specified $2n$ conditions

$$T_k = 0, \quad R_k = 0. \tag{13}$$

In common case, the generalized displacement field X must satisfy $2n$ generalized BCs

$$X_k = X_k^0, \quad d^\perp X_k = X_k^{0'},$$ (14)

or combination of Eqs.13,14 – the impedance BCs or BCs of the third order with X_k^0 and $X_k^{0'}$ as the given quantities. It should be emphasized that the above formulation of the equations and boundary conditions does not depend on the form of the thermodynamic potential w. For the particular case of quadratic dependence of w on the generalized forces, BCs are derived in [7], and for the presence of flexoelectic effect is available in [8]. In the present paper we propose a simple derivation of BCs, which is also suitable for non-linear equations.

Multiferroic layer possessing the flexoelectric effect

As a demonstration of the approach, we consider the problem of the electric and elastic fields distribution in the flexoelectric layer. Denote the thickness of the flexoelectic by L and direct applicate axis perpendicular to the layer. We consider the scalar case, i.e. the displacement vector has one component u_3 depending just on $z = x_3$, and $u_3 = u_3(z)$. For simplicity, the piezoelectric and electrostrictive effects are not taken into account. The derivatives with respect to z will be denoted by primes. Let U be the potential difference at the boundaries of the layer, and assume that these boundaries are stress free. This means that the natural boundary conditions of the form Eq.13 are imposed for the displacement and polarization. In the one-dimensional layer model, the energy density can be written as [2,10]

$$w = \frac{a_1^e}{2} P^2 + \frac{a_2^e}{4} P^4 + \frac{a_3^e}{6} P^6 + \frac{c}{2}(u')^2 - f^{(1)} P u'' - f^{(2)} P' u' + \frac{g}{2}(P')^2 + \frac{h}{2}(u'')^2 +$$
$$P\varphi' + \frac{\varepsilon_0}{2}(\varphi')^2 + \alpha_1 PM + \frac{\alpha_2}{2} P^2 M^2 + \frac{a_1^m}{2} M^2 + \frac{a_2^m}{4} M^4 + \frac{a_3^m}{6} M^6 + M\psi' + \frac{\mu_0}{2}(\psi')^2,$$ (15)

where $a_{1,2,3}^{e,m}$ are the electric and magnetic coefficients of the Ginzburg-Landau theory, c module of elasticity, $f^{(1,2)}$ flexoelectric coefficient, $\{g,h\}$ gradient coefficients, $\{\varepsilon_0, \mu_0\}$ electric and magnetic constants, $\alpha_{1,2}$ linear and quadratic magnetoelectric coefficients. The variation of Eq. 15 with respect to $\{P, M, u, \varphi, \psi\}$ leads to the following equations

$$E = a_1^e P + a_2^e P^3 + a_3^e P^5 + \alpha_1 M + \alpha_2 PM^2 - fu'' - gP'',$$
$$H = a_1^m M + a_2^m M^3 + a_3^m M^5 + \alpha_1 P + \alpha_2 P^2 M,$$
$$D = \varepsilon_0 E + P, \quad B = \mu_0 H + M, \quad \varsigma_u = cu' + fP' - hu''',$$
$$D' = 0, \quad B' = 0, \quad \varsigma_u' = 0.$$ (16)

Here, D and B are the electric and magnetic inductions. The boundary conditions arising from Eq. 15 are

$$\varphi\big|_{z=0} = 0, \quad \varphi\big|_{z=L} = U, \quad B\big|_{z=0,L} = 0, \quad \varsigma_u\big|_{z=0,L} = 0, \quad \varsigma_P\big|_{z=0,L} = 0, \quad \sigma\big|_{z=0,L} = 0.$$ (17)

The quantities

$$\sigma = cu' - f^{(2)}P, \quad \varsigma_u = cu' + fP' - hu''', \quad \varsigma_P = gP' - f^{(2)}u', \quad \tau = hu'' - f^{(1)}P,$$ (18)

are the defined above conventional, generalized, and higher stresses. Thus, the total number of BCs for the flexoelectric layer is ten. For the investigation of the interfaces between the multiferroic layers, we must add another ten continuity conditions for the quantities

$$\{\varphi, \psi, D, B, P, u, \sigma, \varsigma_u, \varsigma_P, \tau\}.$$ (19)

The resulting Eqs. 16 and boundary conditions Eqs. 17,19 completely describe the electro-magneto-elastic field in flexoelectric layers.

The normal derivatives have to be continuous in polarization gradient theory. If not, the square of the delta function arises in the thermodynamic potential, which becomes unbounded. Thus, the minimum condition of the thermodynamic potential becomes meaningless. In particular, it was explicitly demonstrated in [11]. The situation is similar in the conventional electrical engineering. The continuity of the electric potential at the interfaces is required simultaneously with the continuity of the normal component of the electric displacement. The latter is proportional to the derivative of the potential along the normal.

Summary
We have presented a theoretical framework for describing and evaluating thermodynamics of finite-size multiferroic systems. A special attention has been given to the exact formulation of the boundary conditions at the interface of the composites. For this purpose, the extreme thermodynamic potential condition has been used. It was assumed that this potential is determined by integrating over the entire volume. The mathematically rigorous investigation on the basis of the calculus of variations gave us a possibility to write down a complete set of boundary conditions for multiferroics, accounting for the deformation and polarization gradients. As an example of the model application, we have studied the multiferroic layer having flexoelectric effect. Note that the demonstrated results can be easily generalized to the case of cylindrical and spherical boundaries.

Acknowledgements
A.A. Amirov acknowledges to Russian Science Foundation (project № 18-79-10176) for financial support part of theoretical studies. Contribution of A.S. Starkov was supported by RSCF, research project No.18-19-00512.

References
[1] L.D. Landau, E.M. Lifchits, L.P. Pitaevski, Course of theoretical physics: theory of elasticity, Pergamon, New York, 1981.

[2] P.V. Yudin, A.K. Tagantsev, Fundamentals of flexoelectricity in solids, Nanotechnology 24 (2013) 432001. https://doi.org/10.1088/0957-4484/24/43/432001

[3] I.A. Starkov, A.S. Starkov, Modeling of efficient solid-state cooler on layered multiferroics, IEEE T. Ultrason. Ferr. 61(2014) 1357-1363. https://doi.org/10.1109/TUFFC.2014.3043

[4] A.K. Tagantsev, A.S. Yurkov, Flexoelectric effect in finite samples, J. Appl. Phys. 112 (2012) 044103. https://doi.org/10.1063/1.4745037

[5] V.M. Agranovich, V.L Ginzburg, Spatial dispersion in crystal optics and the theory of excitons (Vol. 18), Interscience Publishers, New Delhi, 1966.

[6] M. Gharbi, Z.H. Sun, P. Sharma, K. White, S. El-Borgi, Flexoelectric properties of ferroelectrics and the nanoindentation size-effect, Int. J. Solid Struct. 48 (2011) 249-256. https://doi.org/10.1016/j.ijsolstr.2010.09.021

[7] R.D. Mindlin, Second gradient of strain and surface-tension in linear elasticity, Int. J. Solid Struct. 1 (1965) 417-438. https://doi.org/10.1016/0020-7683(65)90006-5

[8] A.S. Yurkov, Elastic boundary conditions in the presence of the flexoelectric effect, JETP Letters 94 (2011) 455-458. https://doi.org/10.1134/S0021364011180160

[9] L.E. Elsgolts, Differential equations and variational calculus, Nauka, Moscow, 1969.

Shape Memory Alloys – SMA 2018   Materials Research Forum LLC
Materials Research Proceedings **9** (2018) 167-173   doi: http://dx.doi.org/10.21741/9781644900017-32

[10] A.S. Starkov, O.V. Pakhomov, I.A. Starkov, Theoretical model for thin ferroelectric films and the multilayer structures based on them, J. Exp. Theor. Phys.+ 116 (2013) 987-994. https://doi.org/10.1134/S1063776113060149

[11] S.P. Zubko, N.Y. Medvedeva, The size effect in a layered ferroelectric structure. Tech. Phys. Lett.+ 40 (2014) 465-467. https://doi.org/10.1134/S1063785014060145

Shape Memory Alloys – SMA 2018
Materials Research Proceedings **9** (2018) 174-177

Materials Research Forum LLC
doi: http://dx.doi.org/10.21741/9781644900017-33

Structure and Martensitic Transformations of Hybrid sp^2+sp^3 Carbon Phases

Maksim I. Tingaev[1,a*], Evgeny A. Belenkov[1,b]

[1]Chelyabinsk State University, 129 Brat'ev Kashirinykh Str., Chelyabinsk 454001, Russia

[a]tingaevmi@yandex.ru, [b]belenkov@csu.ru

*corresponding author

Keywords: Crystal Structure, Phase Transitions, Carbon Compounds, Ab Initio Calculations

Abstract. The geometrically optimized structures of twenty-two hybrid $sp^2+ sp^3$ carbon phases modelled of L_6, L_{4-8}, L_{3-12} or L_{4-6-12} graphene layers was calculated by the molecular mechanics method (MM+). In these compounds, all three- and four-coordinated atoms are in equivalent structural positions. Crystal phase lattices are hexagonal, orthorhombic, triclinic or monoclinic symmetry. The ratio of the number of sp^3-hybridized atoms to the number of atoms in the state of sp^2 hybridization in these phases varies from 0.5 to 3. After geometrical optimization by the density functional theory method (DFT), the structures of most phases were transformed into the structures of the graphene layers or the structure of diamond-like phases.

Introduction

Carbon atoms in compounds may be in different structural states with different coordination which depends on the number of nearest neighboring atoms [1,2]. There are a few hybridizations of the electron orbitals in atoms depending on the coordination: two-coordinated states correspond to sp hybridization, three-coordinated – sp^2, and four-coordinated – sp^3. All allotropic carbon forms such as carbine, graphene and diamond, have atoms in the same structural and hybridized states – sp, sp^2 or sp^3, respectively. Due to the strong differences in the electronic configuration, the properties of the allotropic forms differ significantly. It is also possible to have hybrid carbon materials consisting of carbon atoms in various hybridized states $sp+sp^2$, $sp+sp^3$, sp^2+sp^3 and $sp+sp^2+sp^3$. The properties of such materials vary in wide ranges depending on the ratio of atoms in different hybridized states.

The most interesting materials are the sp^2+sp^3 carbon phases, which must have a solid three-dimensional linked structure, similar to the structure of diamond-like phases [3-8]. Hybrid sp^2+sp^3 materials can be obtained in a result of partial "cross-linking" of precursor nanostructures consisting of carbon atoms in three-coordinated states [9-11]. Carbon nanostructures with $0D_c$, $1D_c$ or $2D_c$ crystallographic dimensions [1,2] can be used as precursors. Such nanostructures are: fullerene-like clusters belonging to the structural group $[0D_c, 3]$; carbon nanotubes belong to structural group $[1D_c, 3]$ and graphene layers are a structural group $[2D_c, 3]$. In this paper, we studied hybrid carbon phases that could be formed from layers of the four main polymorphic varieties of graphene.

A model scheme for creating the structure of hybrid sp^2+sp^3 phases from graphene layers

For the construction of hybrid phases, graphene layers of four main structural varieties were selected: L_6, L_{4-8}, L_{3-12}, L_{4-6-12}, in which all carbon atoms are in equivalent structural states. Using these layers, hybrid sp^2+sp^3 phases can be formed with a minimum number of atomic positions (phases of the main structural forms), which should be the most stable of structures of this type.

In this work hybrid sp^2+sp^3 phases from graphene layers were obtained according to the following scheme. Each atom of the graphene layer in three-coordinated (sp^2-hybridized) state

Published under license by Materials Research Forum LLC.

Shape Memory Alloys – SMA 2018   Materials Research Forum LLC
Materials Research Proceedings 9 (2018) 174-177   doi: http://dx.doi.org/10.21741/9781644900017-33

can form an additional covalent bond and pass to the sp^3-hybridized state. An additional bond can be formed with atoms of the bordering graphene layer. Some of the atoms of the layers were cross-linked with atoms of bordering layers. Each layer in the stack (in original cluster) has two bordering layers, the top and the bottom. Therefore, each atom in a layer can form an additional link to either the top or bottom layer. To describe the procedure of cross-linking of the layers we denote atoms, sew with the upper layer, the symbol "A", and the atoms linked with the lower layer – "B". Some of the atoms in the original graphene layers should have no link with the atoms of neighboring layers. Such atoms sign as "C" symbol.

As a result of theoretical analysis, it is possible to consider all variants of cross-linking and to describe the structure of various hybrid phases, which are modeled by this method. The number of resulting structural varieties in this case is not limited, but the most interesting are the structures of sp^2+sp^3 phases of atoms with only two different crystallographic equivalent states. These structures should be the most stable and they can be modeled only in strictly periodic order of cross-linking graphene layers.

The structure of the graphene layers and hybrid phases in the basis set of elementary cells. For periodic cross-linking of the hybrid phases is necessary to specify the order of cross-linking within one cell and repeat it to the entire layer with periodic broadcasts. The size of the unit cell in the layer must be chosen so there at least three atoms. Only in this case we can cross-link up and down. The analysis of different variants of cross-linking showed that using graphene layers L_6 it is possible to construct 8 phases, for graphene layers L_{4-8} – 4 phases, from layers L_{3-12} – two phases, from layers L_{4-6-12} – eight phases.

Results of calculations of structure of sp^2+sp^3 hybrid phases

As a result of calculations performed by molecular mechanics MM+ found that the structure of the phases is significantly different from each other. In the structure of phases there are tubular channels of different cross-sections. In two-dimensional projections of the phases structure there are sections of four-, six- and octagonal type, where the scheme in the section is formed by interatomic bonds. Examples of the structure of four hybrid phases are shown in Fig.1.

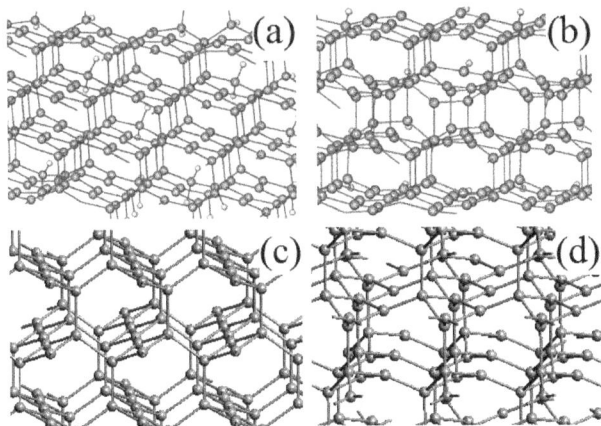

Figure 1. Geometrically optimized hybrid carbon phases: (a) L_6a3; (b) L_6a4; (c) $L_{4-8}a1$; (d) $L_{4-8}a2$.

In the central part of the clusters, calculated for each of the phases the elementary cells were determined. Central part were taken because of minimal deformations due to the influence of surface effects.We measured lengths of the elementary translation vectors a, b and c, and the

Shape Memory Alloys – SMA 2018   Materials Research Forum LLC
Materials Research Proceedings **9** (2018) 174-177   doi: http://dx.doi.org/10.21741/9781644900017-33

angles between them α, β and γ. The crystal lattices of hybrid sp^2+sp^3 phases belong to hexagonal, orthorhombic, monoclinic and triclinic symmetry. The elementary cells contain from 6 to 36 atoms. It were also found values of the structural parameters. Density of the calculated hybrid phases varies in the range of 2.97 g/cm^3 to 3.65 g/cm^3. The density of most hybrid sp^2+sp^3 phases is less than that of diamond (3.4 g/cm^3 [12]) and higher than that of graphite (2.2 g/cm^3 [12]). For L_6a5 phase density is equal to 3.65 g/cm^3, which is higher than the density of diamond. The loosest structure with a minimum density of 2.97 g/cm^3 is observed for the L_6a7 phase. In the constructed phases there are two different structural positions of atoms, which corresponds to the three- and four-coordinated arrangement of atoms (sp^2 or sp^3 hybridized state of electronic orbitals).

In calculations by the DFT method in the generalized gradient approximation (GGA), the structure of almost all hybrid phases transformed into the structure of diamond-like or graphene-like phases consisting only of sp^3 or sp^2 hybridized atoms. Such structural transformations seem to occur in the carbon materials subjected to the treatment of high pressures. Effect of high pressures on bundle of graphene layers causes the process of cross-linking. Hybrid sp^2+sp^3 phases can be formed in case cross-linking is not finished. However, as soon as the external pressure stops, the transformation of the structure begins and can return to its original state in the form of a stack of graphene layers or in the structure of diamond-like phases. In the process of such structural transformations, atoms move at distances smaller than interatomic ones, so such phase transitions can be considered as martensitic.

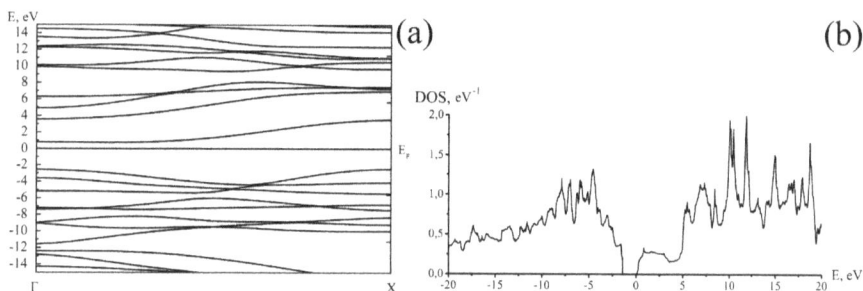

Figure 2. Band structure (a) and density of electronic states (b) of L_6a3 phase.

The only one hybrid phase whose structure does not collapse during optimization is the L_6a3 phase. The model structure of this phase was constructed by merging neighboring graphene layers in the sequence of CAACBB. The sublimation energy of this phase (7.38 eV/atom) has a maximum value in comparison with the sublimation energies of the other studied hybrid phases. The width of the band gap at the level of the Fermi energy for this phase is 1.43 eV (Fig. 2). Thus, this phase should have semiconductor properties.

Summary

Thus, it was found that with partial cross-linking of graphene layers, it is possible to model the structure of 22 sp^2+sp^3 hybrid carbon phases. Using the DFT-GGA method, the structure of most of these phases transformes into the graphene layers or diamond-like phases. The only one hybrid phase whose structure does not collapse during optimization is the L_6a3 phase. The partial cross-linking of graphene layers considered in the work can occur when external pressures of $\sim$ 10–30 GPa are applied to graphite. However, the hybrid phases that form during this process will be metastable and, when the external pressure is removed, will turn into graphite or diamond.

Shape Memory Alloys – SMA 2018   Materials Research Forum LLC
Materials Research Proceedings **9** (2018) 174-177   doi: http://dx.doi.org/10.21741/9781644900017-33

References

[1] E.A. Belenkov, V.A. Greshnyakov, Classification of structural modifications of carbon, Physics of the Solid State. 55 (2013) 1754-1764. https://doi.org/10.1134/S1063783413080039

[2] E.A. Belenkov, V.A. Greshnyakov, Classification schemes of carbon phases and nanostructures, New Carbon Materials. 28 (2013) 273-283. https://doi.org/10.1016/S1872-5805(13)60081-5

[3] M. Nunez-Regueiro, P. Monceau, A. Rassat, P. Bernier, A. Zahab,Absence of a metallic phase at high pressures in C_{60}, Nature. 354 (1991) 289-291. https://doi.org/10.1038/354289a0

[4] V.V. Brazhkin, A.G. Lyapin, Yu.V. Antonov, Amorphization of fullerite (C_{60}) at high pressures, JETP Lett. 62 (1995) 350-354.

[5] M.J. Bucknum, R. Hoffmann, Hypothetical dense 3,4-connected carbon net and related B_2C and CN_2 nets built from 1,4-cyclohexadienoid units, J. Am. Chem. Soc. 116 (1994) 11456-11464. https://doi.org/10.1021/ja00104a027

[6] A. Kuc, G. Seifert, Hexagon-preserving carbon foams: Properties of hypothetical carbon allotropes, Phys. Rev. B. 74 (2006) 214104. https://doi.org/10.1103/PhysRevB.74.214104

[7] E.A. Belenkov, A.L. Ivanovskii, S.N Ul'yanov, F.K. Shabiev, New framework nanostructures of carbon atoms in sp^2 and sp^3 hybridized states, J. Str. Chem. 46 (2005) 961-967. https://doi.org/10.1007/s10947-006-0228-5

[8] N. Park, K. Park, M. H. Lee, J. Ihm, Electronic structure and mechanical properties of graphitic triclinic and honeycomb lattices, J. Korean Phys. Soc. 37 (2000) 129-133.

[9] E.A. Belenkov, M.I. Tingaev,Structure of new sp^2+sp^3 hybrid carbon phases by means of alignmenting of armchair single-walled carbon nanotubes, Letters on Materials 5 (2015) 15-19. https://doi.org/10.22226/2410-3535-2015-1-15-19

[10] M.I. Tingaev, E.A. Belenkov, Hybrid sp^2+sp^3 carbon phases created from carbon nanotubes, J. Phys. Conf. Ser., 917 (2017) 032013. https://doi.org/10.1088/1742-6596/917/3/032013

[11] M.I. Tingayev, V.M. Berezin, E.A. Belenkov, Computation of structure and electronic properties of hybrid phase formed by polymerization of C_{20} fullerenes, Chelyabinsk Physical and Mathematical Journal, 2 (2017) 489-496.

[12] H.O. Pierson, Handbook of carbon, graphite, diamond, and fullerenes: properties, processing and applications, Noyes Publications: Park Ridge, New Jersey, USA,1993.

Keyword Index

Author Index

About the Editors

Professor Dr. Vasiliy D. Buchelnikov

Institution: Chelyabinsk State University
Department: Condensed Matter Physics
Address: Chelyabinsk State University, 129 Brat'ev
                      Kashirinykh Str., Chelyabinsk 454001, Russia
E-mail: buche@csu.ru

V.D. Buchelnikov graduated from M.V. Lomonosov Moscow State University in 1978. He got his doctoral (Ph.D.) degree in 1983 and higher D.Sci. degree in 1992. His research fields are theoretical investigations of phase transformations, ground state structural and magnetic properties of magnetically ordered shape memory alloys, peculiarities of coupled spin, sound and electromagnetic waves, and damping of magnetoelastic waves in magnetoordered substances near the magnetic phase transitions, microwave heating of metallic powders. Since 1978 Prof. Vasiliy Buchelnikov has more than 300 publications, including internationally refereed journals, reviews, monographs, monograph chapters and presentations given at Russian and International conferences. From 2008 to 2016 he was the Editor-in-chief of Bulletin of Chelyabinsk State University (Physics series). Since 2016 till now Prof. Vasiliy Buchelnikov is Chelyabinsk Physical and Mathematical Journal Editor-in-chief. Nowadays, he works as Vice Rector for Science and Chair of Condensed Matter Physics Department at Chelyabinsk State University.

Dr. Vladimir V. Sokolovskiy

Institution: Chelyabinsk State University
Department: Condensed Matter Physics
Address: Chelyabinsk State University, 129 Brat'ev
                      Kashirinykh Str., Chelyabinsk 454001, Russia
E-mail: vsokolovsky84@mail.ru

V.V. Sokolovskiy graduated from Chelyabinsk State University in 2007. After that he became a PhD student and got his doctoral (Ph.D.) degree in 2010. Despite his young age (33 years), he has obtained the D.Sci. degree in 2017. His research fields are theoretical investigations of phase transformations, ground state structural and magnetic properties of magnetically ordered shape memory alloys. Dr. Vladimir Sokolovskiy was granted several times by the funding of Chelyabinsk State University (2010, 2011, and 2014), Council for Grants of the President of the

Russian Federation (2013 and 2016), Russian Foundation for Basic Research (2012), and Russian Science Foundation (2017). At present, he works as associated professor at Chelyabinsk State University.

Dr. Mikhail A. Zagrebin

Institution: Chelyabinsk State University
Department: Radiophysics and Electronics
Address: Chelyabinsk State University, 129 Brat'ev
                     Kashirinykh Str., Chelyabinsk 454001, Russia
E-mail: miczag@mail.ru

M.A. Zagrebin graduated from Chelyabinsk State University in 2008. After that, he became a Ph.D. student and got his doctoral (Ph.D.) degree in 2009. His research fields are a phenomenological description of phase transformations, ground state structural and magnetic properties of magnetically ordered shape memory alloys. Dr. Mikhail Zagrebin was granted several times by the funding of Chelyabinsk State University (2012 and 2017) and Russian Foundation for Basic Research (2012 and 2014). From 2010 to 2016 he was an executive secretary of Bulletin of Chelyabinsk State University (Physics series). Since 2016 till now Dr. Mikhail Zagrebin is Chelyabinsk Physical and Mathematical Journal executive secretary. Nowadays, he works as associated professor at Chelyabinsk State University.

Olga N. Miroshkina

Institution: Chelyabinsk State University
Department: Condensed Matter Physics
Address: Chelyabinsk State University, 129 Brat'ev
                     Kashirinykh Str., Chelyabinsk 454001, Russia
E-mail: miroshkina.on@yandex.ru

O.N. Miroshkina graduated from Chelyabinsk State University in 2016. After that, she became a Ph.D. student of the Condensed Matter Physics Department. Her research fields are a phenomenological description of phase transformations, ground state structural, magnetic, and vibrational properties of magnetically ordered shape memory alloys. Olga Miroshkina was granted by the funding of Chelyabinsk State University in 2018. At present time, she works as research assistant at Chelyabinsk State University.

www.ingramcontent.com/pod-product-compliance
Lightning Source LLC
Chambersburg PA
CBHW070720220326
41598CB00024BA/3241